HO]

The Case for Hard Thinking, Honesty and Humility when Assessing Environmental Health Risks

E. Joseph Duckett, PE, PhD
Jeffrey L. Pierce, PE

HOLD IT! The Case for Hard Thinking, Honesty and Humility when Assessing Environmental Health Risks

Print ISBN 978-1-949267-81-5
ebook ISBN 978-1-949267-82-2

Front cover illustration by Olha Bondarenko
Cover design by Guy Corp, www.GrafixCorp.com

STAIRWAY≡PRESS

STAIRWAY PRESS—APACHE JUNCTION

www.stairwaypress.com
1000 West Apache Trail, Suite 126
Apache Junction, AZ 85120 USA

Joe's Dedication

In memory of my Mother, Elizabeth (Betty) Duckett, and my Father, Edward Joseph (Ed) Duckett, who gave me both my name and my lifelong interest in learning.

Jeff's Dedication

To my partner in life, my wife Samantha, who with a BS in engineering and an MBA, is able to engage in lively discussions with me on topics like those addressed in this book. Invariably, we find that in these discussions that we are "singing to the choir."

Endorsements

Hold It! is a thoroughly researched and thought-provoking examination of environmental risk assessment. In challenging popular myths, misconceptions, and misinformation about man-made threats to human health and the environment, the authors navigate the hype of both alarmists and deniers. They illuminate a path forward based on research, informed environmental choices, and intellectually honest decision-making.
—Tom O'Toole, University Assistant Vice-Chancellor

The authors have presented a history of what happens when the public doesn't critically consider the assertions of "experts" and those zealously pursuing an outcome at all costs. We had better learn that it's far more effective, economically, and socially, to do things right, once, the first time.
—Vince Brisini, Former Deputy Secretary, Pennsylvania Department of Environmental Protection

Kudos to Duckett and Pierce for cutting through the environmental propaganda on both sides like a hot knife through butter. Through critical and unbiased analysis (the hallmarks of true science), they clearly show the fallacies of the virtue signalers and the deniers. If we really want to make a difference, we need more objective scientists like the authors that would be believed by both camps and an honest approach by politicians, the media, and, to the extent possible, the general public.

—Mäny Emamzadeh, Attorney; Financial Firm Chief Executive Officer

The authors provide a valuable look into a series of environmental "incidents" over many decades pointing out that what often begins as "solutions or improvements" turns out to create significant new, unintended, environmental harms.

The authors take the positive step to offer improvements to our crises—focused, reactive legislative/regulatory system. They suggest the search for the truth must include a periodic reevaluation of priorities, a focus on prevention rather than delayed responsiveness and a realization that the issues and solutions transcend international boundaries.

One can hope that this search for the truth is the beginning of a national dialog.

—Paul King, Former Environmental Citizens Group President and Allegheny County, PA Board of Health Chair

As a professional in the environmental industry for more than 20 years, I found this book fascinating and informative. The authors provide an objective point of view on many controversial environmental topics of our generation (so that the reader is better informed of facts and myths related to these topics).

Before discussing environmental matters with your friends, colleagues, neighbors...HOLD IT! and read this book first.

—Amy Veltri, Environmental Engineer; Chief Executive Officer

Contents

CHAPTER 1: INTRODUCTION

Experts Speak and the Public Listens

IN JANUARY 1970, LIFE Magazine reported that...

> ...*scientists have solid experimental and theoretical evidence that, in one decade* [i.e., by 1980] *urban dwellers will have to wear gas masks to survive air pollution and...by 1985, air pollution will have reduced the amount of sunlight reaching the earth by one-half...* [1]

Source: Life Magazine January 30, 1970; Creative Commons license

In another pronouncement that drew attention, Newsweek Magazine published an April 28, 1975 article under the headline *The Cooling World* that asserted that global temperatures had been falling "with terrible consequences for food production." [2]

The same article claimed that meteorologists "are almost unanimous in the view that the [cooling] trend will reduce agricultural productivity for the rest of the century." A similar prediction, also made in the 1970's, was that the earth would be 4°F colder by 1990 and 11°F colder by 2000. [3]

A prominent environmental scientist, Paul Ehrlich, warned in the May 1970 issue of Audubon that exposure to DDT and other chlorinated hydrocarbons had reduced the life expectancy of Americans born after 1946 to 49 years, and that if the then current pattern continued, life expectancy would be further reduced to 42 years by 1980. [4]

Paul Ehrlich had previously warned in his book *The Population Bomb* (1968) that...

> *...in the 1970's hundreds of millions of people will starve to death despite any crash programs embarked on now.*

Paul Ehrlich is not without credentials. He still holds the position of Bing Professor Emeritus of Population Studies of the Department of Biology of Stanford University, and he is President of the Stanford Center for Conservation Biology.

None of the above predictions were correct.

As troubled as the world has been by the COVID-19 Pandemic, and other longer-term problems, sunlight still reaches the earth, global food production has outstripped population growth, global cooling has not happened, and the average life expectancy in the USA has risen to almost 79 years. **India now produces more than three times as much wheat and rice as it did in the 1970's and its economy has grown by a factor of 50.** [5] The predictions were exaggerations, advanced in the name of environmental protection.

HOLD IT!

In the opposite vein, both tetraethyl lead (TEL) gasoline additives and leaded paints were introduced early in the 20th Century as low risk advancements in automotive engine performance and paint appearance, respectively.

Despite some early cautionary warnings, leaded paint was advertised with the phrase: "lead helps guard your health" [6] and leaded gasoline was described as "posing no risk at all to public health." [7]

Source: dutchboy.com/forever/paint Creative Commons license

Source: dutchboy.com/forever/paint Creative Commons license

Both statements were later shown to be false as the risks of lead exposure, especially among children, became documented. Dismissive statements about environmental risks can be exaggerated just as alarmist statements can be exaggerated.

Purpose of This Book

This book is for realists in pursuit of environmental and human health improvement. It can be considered CPR for environmental panic attacks or a splash of humility to make us think twice before lunging at popular, often simplistic, solutions.

Christine Whitman, former Governor of New Jersey, and former Administrator of the United States Environmental Protection Agency (U.S. EPA) once commented that, when dealing with environmental issues, the middle ground between extreme alarmism and simplistic rejection is the most difficult position to hold.

She was right.

And yet, moderation, even-handedness and genuine rationality are too often absent from our environmental protection narratives. Science too often takes a back seat to sensationalism, ignorance, and greed.

There are many positives about the field of environmental protection. Protecting human health and natural resources are just two. The technologies for environmental protection are imaginative.

The kaleidoscopic mix of science, politics and economics in environmental affairs is fascinating. Every so often, however, it is worthwhile taking time to carefully think through environmental issues before seizing on quick solutions and believing that we have always known the proper solution to environmental problems.

As noted above, one of the decidedly negative aspects of the environmental field is its repeated reliance on exaggeration either to warn of—or dismiss—health threats and remedies.

For many of us, the official start of the environmental movement was in the late 1960's, an era when the only two options for civilization were portrayed as either complete destruction or total utopia. Maybe the use of exaggeration was needed then to compete for public attention against other causes. Maybe it's just because most environmental damage usually takes a long time to become evident, but people only pay attention to threats when they are perceived as being very severe and imminent.

Whatever the reasons, there is no denying that, as Governor Whitman said, the middle ground of environmental protection is difficult to find, maintain and defend. How else to explain the shifting views of worldwide climate change which, over a relatively short 40-year span, have swung from 1970 predictions of worldwide calamity from global <u>cooling</u> to current worldwide calamity from global <u>warming</u>.[8]

What has happened is that public discourse on environmental problems has been taken over by propagandists more interested in advancing (or defeating) a cause than by illuminating truth. Such propaganda is found among many sectors of interest in environmental issues. Industry groups, citizen groups and political parties have all engaged.

It is unfortunately true that, as one author put it...

...the amount of hyped news coverage bears little relation to measurable or calculable risks.[9]

Without question, and especially in the USA, the past several decades have seen significant improvements in most measures of environmental quality: ambient air pollutant concentrations, water quality indices, hazardous waste containment and solid waste recycling, to name a few indicators. Invoking the word "environment" has become an almost automatic trigger towards positive sentiments and unassailable purpose.

Despite such progress, not all activities labeled "environmental" are positive or even beneficial to our world. The word "environment" itself has diminished in meaning as it has been applied to everything from cigarette sales to garbage collection.

This book will expose some of the examples of phony and misguided environmental activities as distinguished from those that generate actual improvement.

One of the hallmarks of environmental issue debates is often the unyielding certainty (orthodoxy?) with which the proponents (one side or the other) present their positions. Most environmental issues are highly nuanced and judgmental enough that neither the issue itself nor its resolution can fit neatly into a sound bite, bumper sticker, or tweet.

Corrections are hard to admit.

There are too few examples of public admissions of policies, positions, or theories that have proven to be wrong.

Mahatma Gandhi was quoted as saying that "science without humility is one of the seven deadly sins." The word "environmental" could easily slide into the start of Gandhi's quote.

HOLD IT!

Source: London Remembers.com Creative Commons license

Along the same theme as Gandhi's quote is this 1915 comment from Nikola Tesla:

> *It is paradoxical, yet true, to say that the more we know the more ignorant we become...for it's only through enlightenment that we become conscious of our limitations.*

More recently (2020), another author wrote that...

> *...despite decades of news media attention, many remain ignorant of basic facts of environmental issues.*[10]

This book is an attempt to stake out a middle zone among a variety of environmental and health issues, both past and present. Its aim is toward realism and truth with a call for humility. It recognizes that environmental threats should neither be exaggerated nor simply dismissed. It is pro-science but recognizes that scientific conclusions must evolve as learning progresses.

In recognizing the evolution of environmental science, we find ourselves in good company. Although never known as an environmental specialist, but always a keen observer of science and human nature, none other than Ben Franklin in the 1700's commented on the sometimes-slow recognition of environmental risks.

Referring to lead (as used in printing presses, plumbing and paints) as a "poison," he wrote in 1786:

> *You will see that this mischievous effect from lead is at least 60 years old and you will observe with concern how long a useful truth may exist and be known before it is generally received and practiced on.* [11]

In the chapters that follow, we address historical misconceptions; notable successes; exaggerated risks; unintended consequences; regulatory overreach; marketing propaganda/gimmickry; myths and distortions. We devote a separate chapter to global climate change, which is clearly the dominant environmental issue of this century.

Finally, we present ideas for re-infusing realism and honesty into environmental decision-making with the goal of improving human health and natural resources rather than promoting special business/political interests.

Throughout this book, examples are drawn from the two allied fields of environmental and health protection. This is because so much of the motivation for improved environmental conditions has stemmed from the drive to improve human health. Another reason for overlapping environmental and health examples is they both require informed judgment calls, some of which have been spot on while others have not. The needs for unrushed judgment and scientific humility apply across the environment-health spectrum.

Most, but not all, of the examples cited here are from the USA. Many examples are drawn from the Western Pennsylvania (Pittsburgh) region where we both grew up, and some from California, where Jeff moved to in the late-1980s. Most of the covered issues also apply to other regions and countries as well.

Here goes...

---

[1] *Ecology Becomes Everybody's Issue*, Life Magazine, January 30, 1970

[2] Gwynne, P., *The Cooling World*, Newsweek, April 28, 1975, p. 64

[3] Waters, H., *Why Didn't the First Earth Day's Predictions come true? It's Complicated*, Smithsonian Magazine, April 22, 2016

[4] Ehrlich, P., Audubon, May 1970

[5] Gates, B, *How to Avoid Climate Disaster*, Alfred A. Knopf, New York, New York, 2021

[6] Peeples, L., *Lead Paint Trial: Did Industry Promote Product Knowing of Its Toxic Dangers*, ENVIRONMENT, July 2013

[7] Davey, K., *Review of Lead Wars*, August 19, 2013

[8] Perry, M.J., *18 Spectacularly Wrong Predictions Made Around the Time of the First Earth Day in 1970*, American Enterprise Institute, April 22, 2018

[9] Mazur, A., *True Warnings and False Alarms*, RFF Press, 2004

[10] Shellenberger, M., *Apocalypse Never*, Harper Collins Books, 2020

[11] Smellie, W.G., *Public Health: Its Promise for the Future*, MacMillan, 1955 as quoted in Vesilind, P.A., Short Case Studies and Incidental Lecture material in Environmental Engineering, Duke University, 1978

E. Joseph Duckett and Jeffrey L. Pierce

CHAPTER 2: HISTORICAL MISCONCEPTIONS

CONCERN FOR THE environment didn't begin in the 1960's or even in the 20th Century. The term "environment" grew popular over the past five decades, but recognition of our environment's influences on health extends at least back to the 1800's.[1,2] Often, this recognition was misguided (badly distorted) as evidenced by some of the popularized environment and health fallacies of the past. In hindsight, many of them are now even comical.

Even in the 1800's, the first stirrings of recognized connections between human health and environmental conditions were happening. John Snow (an English Physician, not the Game of Thrones character), in his now famous epidemiological observation, noted the link between the spread of Cholera disease and the common use of a contaminated water well.[3]

However, scientific reasoning like Snow's did not establish a firm going forward basis for environmental health understanding and action. Alexis deTocqueville in his *Democracy in America* noted that the combined influences of democracy and economic opportunity in the early 1800's put an emphasis on profit-making technology over basic science.[4] Through the mid-1800's, Romanticism (feelings, sensitivity and the supernatural) replaced reason as the dominant influence on understanding the factors affecting human health.

One result of this drift away from science and reason was an abundance of peculiarly odd health practices and environmental misunderstandings.

- People were advised to drink water laced with radium as a cure for arthritis and impotence. When deaths occurred from drinking radioactive water, this practice ceased and yet another connection between drinking water quality and health had been made.[5]

- Bloodletting was considered a treatment for many diseases thought to result from excess fluids in the body.[6]

- A bad case of head lice was thought to be cured by dousing hair with gasoline or kerosene. This may have been effective in killing lice but at the risk of major danger if anyone walked near an open flame or lit a match.[7]

- To relieve the teething pains of children, morphine syrups and/or mercury powders were suggested to well-meaning, but unsuspecting, parents. Similarly, "Cocaine Tooth Drops" were marketed in the late 1800's to provide an "instantaneous cure."[8]

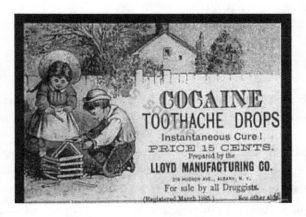

Source: blogspot.com Creative Commons license

In the late 1800's, acid mine drainage, predominantly from coal mines, was considered sufficient to protect natural waters from typhoid and cholera.[9]

- Sulfur dioxide, now recognized as a serious air pollutant, was once thought to provide "cleansing" for people with lung cancer. Cancer victims sometimes traveled to heavily polluted cities to clean their lungs.[10]

- Beyond sulfur dioxide, *smoke* was claimed to kill malaria. Also, to save eyesight because darkened skies reduced glare.[11]

- Smoke was also often associated with wealth. In an 1880 speech by industrialist Robert Ingersoll, he boasted:

    I want the sky to be filled with the smoke of American industry, and upon that cloud of smoke will rest forever the bow of perpetual promise. That is what I am for.[12]

- Rounding out the 1800's, in a tour-de-force of political incorrectness, the smoky air of Pittsburgh in the 1890's was said to account for the "manliness of its men and the absence of frivolity among its women."[13]

Even in the first half of the 20th century, many practices and environment-health understandings that are now debunked were still in vogue.

- Prior to the introduction of penicillin in 1943, patients with syphilis were advised to ingest mercury and/or arsenic. Another treatment was to deliberately inoculate a patient with malaria which would presumably raise body temperature, kill the syphilis and effectively replace it with malaria.[14]

- As a cure for tuberculosis in the early 20th Century, a treatment termed *plombage* was employed. By creating a

14

cavity in a patient's lower lung and filling it with inert material (Lucite ping-pong balls were sometimes used), the aim was to deliberately collapse the upper lung. Strange as it now seems, the theory was that a collapsed lung would heal itself.[15]

- "Sanitized" tape worms were promoted as a way to "always stay thin" and to "banish fat." [16]

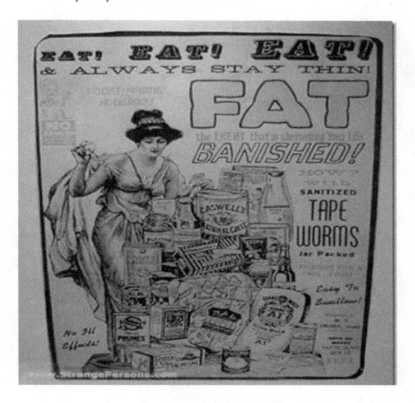

Source: The Lifetree.com Creative Commons license

- A treatment for chronic migraines in the early 1900's was immersion in a warm hydroelectric bath. Ideally, the electric current was kept low enough to avoid electrocuting the patient.[17]

- For young mothers, a case of beer was advertised under the banner "How Mother and Baby Picked Up." The ad claimed that the "malt in beer supplies nourishing qualities that are essential...and the hops act as an appetizing, stimulating tonic." [18] The ad went on to note that "obviously baby participates in its benefits."

Source: blogspot.com Creative Commons license

- An American physician declared that the introduction of the automobile would reduce disease transmission. His theory was that, by replacing horse-drawn carriages, horse manure and flies would no longer "pollute" the streets. His quote: "A serious channel of infection will be done away with, and many lives will be spared. The horseless carriage will greatly reduce the death rate in cities." [19]

Just as some environmental misconceptions persisted into the 1900's, several environmental conditions were starting to be recognized as health threats. Prominent among these were lead, asbestos, sulfur dioxide, "smoke" and drinking water-borne pathogens (especially typhoid bacteria). All of these pollutants are discussed later in this book. Most of them have been very successfully addressed and diminished as health threats…at least in the USA.

Although not entirely an environmental issue, the unfolding of evidence linking inhalation of cigarette smoke and respiratory (and other) diseases somewhat parallels the evolution of understandings about air pollution and health. At least through the 1950's, cigarette smoking was promoted as not only being harmless but, in some instances, actually beneficial. Some examples…

Source: flickr.com Creative Commons license

- Viceroys were promoted with an ad suggesting: "As your Dentist, I would recommend Viceroys."

- An ad for Camel cigarettes pronounced: "More Doctors Smoke Camels than any other cigarette!"

Source: flickr.com Creative Commons license

- Kent cigarettes advertised that, because of their "Micronite" filters, "more scientists and educators" smoke them. The Micronite filter innovation backfired when it was revealed that the filters contained crocidolite asbestos, an environmental hazard in its own right.

18

Source: flickr.com Creative Commons license

Just as cigarette smoke is now fully recognized as a health threat, so also have many more generalized environmental pollutants. Similarly, regulations now govern both cigarette smoking and environmental discharges.

Some environmental protections have worked well, others not so well. Later chapters discuss both notable achievements and missteps.

To conclude this chapter on past misconceptions, an important takeaway is that scientists, even environmental scientists, are not always correct in assessing health risks and remedies. If nothing else, recital of some misguided practices of the past raises cautionary humility warnings about today's prevailing environmental "certainties."

Not all answers are immediately obvious, and most are not simple.

---

[1] Ivey, L., *Smoke Gets in Our Eyes*, Sustainability in History, June 1, 2013

[2] Sebak, R., *This Brit Fell in Love with Pittsburgh's Dirt*, Inside Outside Southside, June 16, 2019

[3] Barry, J.M., *The Great Influenza*, Penguin Books, 2005

[4] deTocqueville, A., Democracy in America, 1863

[5] Tanner, C., *The Quackery Cures of Yesteryear*, The Daily Mail, October 17, 2017

[6] Floyd, B., *From Quackery to Bacteriology: The Emergence of Modern Medicine in 19th Century America: An Exhibition*, University of Toledo, OH, 2019

[7] *9 Terrifying Medical Treatments from 1900 and Their Safer Modern Versions*, Mental Floss, May 18, 2018

[8] *Healthcare in the Late 1800's – A Fun Look at Some Crazy Patent Medicines*, CORE Higher Education Group, 2015

[9] Tarr, J., *Devastation & Renewal*, University of Pittsburgh Press, 2005

[10] Davidson, C., *Air Pollution in Pittsburgh: A Historical Perspective*, Journal at the Air Pollution Control Association, 1979

[11] Westman, R., *Air Pollution in Pittsburgh: A History*, Pittsburgh Engineer, Winter 2006

[12] Hodges, L. *Environmental Pollution*, Holt, Rinehart & Winston, 1977

[13] Gugliotta, A., Class, *Gender and Coal Smoke*, Environmental History, Vol. 5, No. 2 Oxford University Press, 2000

[14] Potter, M., *Health Misconceptions of the 18th Century*, 18th Century History – The Age of Reason and Change, 2020 @ www.history1700s.com

---

[15] *9 Terrifying Medical Treatments from 1900 and Their Safer Modern Versions*, Mental Floss, May 18, 2018

[16] *Healthcare in the Late 1800's – A Fun Look at Some Crazy Patent Medicines*, CORE Higher Education Group, 2015

[17] Tarr, J., *Devastation & Renewal*, University of Pittsburgh Press, 2005

[18] *Pick Up with Blatz Beer*, www.flickr.com/photos

[19] Jay, A. and Frost, D, *The English*, Stein & Day, 1968

CHAPTER 3: NOTABLE SUCCESSES

ESPECIALLY OVER THE past half-century, at least within the USA, very significant successes have occurred in recognizing and controlling environmental pollutants.

Among the more notable successes have been in controlling tetraethyl lead; water-borne typhoid fever; sulfur dioxide; asbestos; and a variety of other targeted air and water pollutants.

Tetraethyl Lead

As long ago as the 1920's, tetraethyl lead (TEL) was added to gasoline as an octane booster—to prevent engine knocking.

Within only a few years after starting this practice, there were suspicions about the health effects of emitting lead compounds in automobile exhaust.[1]

HOLD IT!

Despite several decades of research and evidence that the use of TEL was leading to elevated airborne lead levels and causing a public health problem, it was not until 1973 that the U.S. EPA initiated a phased reduction in the use of TEL, and then in 1996 completely banned TEL from gasoline for all on-road vehicles.[2]

As with many environmental regulations, the TEL ban produced both the intended beneficial result (lower atmospheric lead levels), but then produced an unintended adverse environmental result.

The lowering of atmospheric lead levels has been dramatic (see Figure 1). It has been so successful that lead inhalation, once the subject of popular concern, is hardly even on the environmental radar screen. Average ambient USA concentrations of atmospheric lead decreased more than 98% from 1980 to 2019 and the trend has continued.

Concerns are still raised about occupational health, or very localized exposures to airborne lead, or to drinking water lead contamination (see Chapter 4); but there is no longer a concern for general ambient air lead exposure.

Figure 1—Lead Air Quality, 1980-2019
(Annual Maximum 3-Month Average)
National Trend based on 7 Sites
Source: U.S. EPA

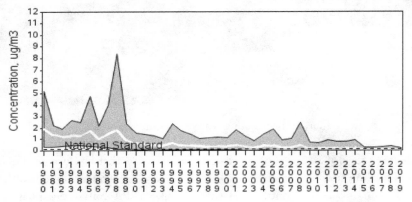

Lead Air Quality, 1980 - 2019
(Annual Maximum 3-Month Average)
National Trend based on 7 Sites

1980 to 2019 : 98% decrease in National Average

Replacements for TEL in gasoline have been effective, for the most part, in preventing excessive engine knocking. However, as is often the case in a response to an environmental improvement mandate, there were unintended consequences. In this case, an adverse consequence from using Methyl Tertiary Butyl Ether (MTBE), as the initial TEL replacement. The consequence, a groundwater impact, is discussed in Chapter 5.

The U.S. EPA's program to phase out TEL was accelerated by another adverse impact of the presence of lead in gasoline. U.S. EPA required that catalytic converters be installed on automobiles sold beginning with the 1975 model year. The catalytic converters were intended to reduce automotive emissions of carbon monoxide (CO) and volatile organic compounds (VOCs). VOCs are precursors to ozone formation (smog). Catalytic converters complete the combustion of VOCs and CO before they became tailpipe emissions. Lead (from TEL) poisons the catalysts, reducing their effectiveness in controlling VOCs and CO. In 1974, U.S. EPA required that all gasoline stations provide at least one grade of unleaded gasoline.

So, even if lead was not an air quality public health concern in itself, lead would have become an air quality public health concern, since lead became an obstacle to implementing the chosen method of solving another air quality public health concern. At the end of the day, the elimination of leaded gasoline not only reduced automotive lead emissions, but it also facilitated automotive VOC and CO emission control.[3]

Typhoid Fever and Drinking Water

At the turn of the 20th century, Pittsburgh, PA had the highest typhoid fever rate in the USA.[4] Typhoid cases were traced to the public drinking water supply.

After bench scale research, engineering design and governmental fiscal commitments, the Pittsburgh Water Works was retrofitted with a new water treatment process, slow-sand filtration. This was new technology at the time (circa 1905) and it was an innovative step for any city.

Within a few years of installing the new filtration equipment (by 1908), Pittsburgh was transformed from having one of the highest typhoid fever rates to among the lowest in the USA. Slow sand filtration continues in use today around the world. Typhoid is no longer considered a significant threat among USA cities with advanced potable water filtration and disinfection systems.

Sulfur Dioxide (SO₂)

For at least three-quarters of a century, sulfur dioxide (SO_2) has been recognized as a serious air pollutant. Inhalation of SO_2, in combination with airborne particulate emissions, has been associated with several respiratory health effects.

Coal combustion has been the major source of SO_2 emissions. In 1980, the average annual USA ambient air concentration of SO_2 was 160 parts per billion (ppb). By 2019, this average was reduced to approximately 10 ppb, a 92% reduction over 40 years. Over this period, the reduction in SO_2 concentrations was continuous.[5]

Figure 2—SO_2 Air Quality, 1980-2019
(Annual 99th Percentile of Daily Max 1-HourAverage)
National Trend based on 35 Sites

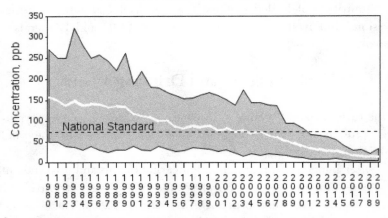

1980 to 2019 : 92% decrease in National Average
Source: U.S. EPA

There are at least three major factors accounting for such dramatic reductions in ambient SO_2. First is the conversion from coal-firing to natural gas-firing—a conversion that has been adopted at many power plants and boilers. Unlike coal, natural gas contains no sulfur, and when burned produces no SO_2. It is reported that from 2008 through 2019, the natural gas share of electric power production doubled to about 40% while the coal-fired share was cut to approximately 36%.[6]

A second major factor has been the downward ratcheting of allowable SO_2 emissions from coal-fired combustion units. Tightened regulations have led to the retrofitting of SO_2 scrubbers to reduce SO_2 emissions for coal-fired combustion units—both power plant boilers and other sources. Over the period 1990-2016, approximately 81% of power production facilities were equipped with such scrubbers.[7]

A third factor was the creation of so-called "cap-and-trade" programs for SO_2. Through such programs, the combined allowable

SO_2 emissions among many sources (primarily power plants) were capped and then further lowered via downward-ratcheted reductions over time. (See Chapter 10 for more on cap-and-trade).

Source: Purdue University Energy Center; Creative Commons license

As with many responses to environmental regulatory mandates, there were some adverse collateral effects of SO_2 reduction. One such effect was due to the use of limestone (CaCO3) as the reagent for SO_2 scrubbing. Limestone is used either directly as the reagent or indirectly after conversion to lime. Either way, 44 tons of carbon dioxide (CO_2) are released to the atmosphere for each 100 tons of limestone consumed. CO_2 has been identified as the most prominent greenhouse gas affecting global climate change. There are other reagents that can be used for SO_2 scrubbing but, for reasons of cost and availability, limestone and lime are by far the predominant SO_2 scrubbing reagents.

One interesting twist to the SO_2 reduction story is the widescale adoption of hydraulic fracking for natural gas production. Due to advancements in horizontal drilling and fracking, the cost and availability of natural gas have improved dramatically in recent years. So much so that natural gas-firing has become very cost-competitive with coal-firing. In turn, many coal-fired boilers have been converted to natural gas-firing, or they have been shut down and replaced with natural gas fired units, both because of the competitive cost advantage of natural gas, and because natural gas-firing does not require the installation and operation of expensive SO_2 scrubbers.

Ironically, there are now environmental pressures to restrain fracking, even though the result of fracking has been an unquestionably sharp downturn in SO_2 (and other) emissions from coal-firing.[8]

Asbestos

Probably no environmental hazard has been recognized for so long and has been addressed so aggressively, yet still remains problematic, as asbestos.

The use of asbestos dates as far back as 4,500 years ago. People made pots and other cooking vessels from it. The word "asbestos" derives from the Greek term for inextinguishable, owing to the heat resistance of asbestos fibers. Chrysotile, one of the principal forms of asbestos, is Greek for "golden fiber," reflecting the high value placed on this seemingly indestructible mineral fiber.[9]

The environmental health risk of asbestos was first documented around 100 A.D. by the ancient Roman scholar, Pliny the Younger.

He observed that quarry slaves who mined asbestos were prone to become ill and should therefore not be purchased.

Fast forward almost 2,000 years and similar observations were made of shipbuilders, miners, insulation installers and other asbestos workers in the early 1900's. By 1930, asbestosis had been recognized as a disease due to asbestos exposure. The discoverer was Dr. E. Merewether, a researcher who recommended that asbestos exposures should be controlled by improved workplace ventilation and the use of masks.[10] He noted that, unlike other acutely toxic worker exposures, asbestosis was a

Source: Clark, C., Faith Explained.6/14/16
Creative Commons license

disease of latency. The effects of asbestos exposure might not show up for many years, even decades. Although coming two millennia after Pliny, Dr. Merewether was still ahead of his time.

From the 1930's through the 1960's, evidence was mounting not only about asbestosis but also lung cancers among asbestos workers. Insurance policies began decreasing coverage and benefits for asbestos workers. A 1960 study confirmed that exposure to asbestos causes mesothelioma (cancer of the membrane lining the lungs) and that the risks could even be transmitted to the children and spouses of workers.

Throughout this period, asbestos use continued to be touted because of its durability, insulating properties and resistance to fire or chemical attack. Asbestos was indeed considered very useful and valuable.

In the 1970's, serious attention began to be paid to asbestos exposures. Shipyard workers from as far back as World War II began developing asbestosis. Major epidemiological studies pinpointed the links between exposures and respiratory diseases.[11]

In 1971, the U.S. Occupational Safety and Health Administration (OSHA) began regulating asbestos exposures. Subsequent regulations of the U.S. EPA and the Consumer Product Safety Commission further tightened the rules for exposures. Lawsuits against asbestos manufacturers and users added even more pressure.[12]

As an almost unique aspect of asbestos protection, regulations and governmental directives not only dealt with new asbestos applications but also pressed for safe removal of existing applications, some of which were decades old. For example, the removal, with containment, of asbestos fireproofing and insulation from schools and public buildings became its own new business, which continues today.

The ironic twist to the asbestos story is that, although many asbestos uses have been curtailed and previous applications have been removed, the actual rates of asbestosis cases are still increasing.[13] This is due to the long latency periods for asbestos-related diseases and to sharpened diagnoses of such diseases.[14]

So, certainly there has been much success with asbestos protection...but with more progress still remaining to be made.

E. Joseph Duckett and Jeffrey L. Pierce

Criteria Air Pollutants

The Clean Air Act of 1970 amended the 1963 Clean Air Act and introduced the term "Criteria Air Pollutants" into the environmental lexicon. Criteria Air Pollutants were considered to be the six most important air contaminants related to human health. They are carbon monoxide (CO); lead (Pb); nitrogen dioxide (NO_2); ozone (O_3), particulate matter (PM) and sulfur dioxide (SO_2). These air contaminants were the pollutants for which there were demonstrated experimental and epidemiological evidence of health effects, as documented in so-called "Air Quality Criteria."

As a result of a decades-long series of actions—legislative amendments (principally in 1977 and 1990), regulatory implementation, downward reductions in standards (re-targeting of goals), technological innovations and fuel switches—dramatic reductions in all of these Criteria Air Pollutants have occurred.[15]

One factor in these reductions has been the re-targeting of ambient air quality goals over time. "National Ambient Air Quality Standards" (NAAQS) are the defined targeted atmospheric concentrations of pollutants. The NAAQS were initially based on the above referenced Air Quality Criteria documents. NAAQS are primarily intended to protect human health. The NAAQS are periodically re-evaluated based on up-to-date research on the relationships between pollutant inhalation and health.[16]

Over time, the NAAQS for criteria pollutants have been re-targeted to lower levels. Table 1 illustrates the progressive re-targeting of ambient air particulate standards. State and local regulations restricting particulate air emissions from individual sources have been correspondingly revised downward, as needed, by air quality management agencies, chasing the lower ambient air quality standards.

The result has been reduced ambient air pollutant concentrations. In some regions, this moving of the "goal posts" may have led to a region not being in compliance for a particular pollutant, giving the impression of lack of progress, in that region, in its air quality improvement.

Table 1—History of the National Ambient Air Quality Standards for Particulate Matter During the Period 1971-2012

Final Rule	Primary/Secondary	Indicator	Averaging Time	Level[1]	Form
1971 36 FR 8186 Apr 30, 1971	Primary	TSP[2]	24-hour	260 µg/m³	Not to be exceeded more than once per year
			Annual	75 µg/m³	Annual geometric mean
	Secondary	TSP	24-hour	150 µg/m³	Not to be exceeded more than once per year
			Annual	60 µg/m³	Annual geometric mean
1987 52 FR 24634 Jul 1, 1987	Primary and Secondary	PM₁₀	24-hour	150 µg/m³	Not to be exceeded more than once per year on average over a 3-year period
			Annual	50 µg/m³	Annual arithmetic mean, averaged over 3 years
1997 62 FR 38652 Jul 18, 1997	Primary and Secondary	PM₂.₅	24-hour	65 µg/m³	98th percentile, averaged over 3 years
			Annual	15.0 µg/m³	Annual arithmetic mean, averaged over 3 years
		PM₁₀	24-hour	150 µg/m³	Initially promulgated 99th percentile, averaged over 3 years; when 1997 standards for PM₁₀ were vacated, the form of 1987 standards remained in place (not to be exceeded more than once per year on average over a 3-year period)
			Annual	50 µg/m³	Annual arithmetic mean, averaged over 3 years
2006 71 FR 61144 Oct 17, 2006	Primary and Secondary	PM₂.₅	24-hour	35 µg/m³	98th percentile, averaged over 3 years
			Annual	15.0 µg/m³	Annual arithmetic mean, averaged over 3 years[3]
		PM₁₀	24-hour[3]	150 µg/m³	Not to be exceeded more than once per year on average over a 3-year period
2012 78 FR 3086 Jan 15, 2013	Primary	PM₂.₅	Annual	12.0 µg/m³	Annual arithmetic mean, averaged over 3 years[3]
	Secondary	PM₂.₅	Annual	15.0 µg/m³	Annual arithmetic mean, averaged over 3 years[3]
	Primary and Secondary		24-hour	35 µg/m³	98th percentile, averaged over 3 years
	Primary and Secondary	PM₁₀	24-hour	150 µg/m³	Not to be exceeded more than once per year on average over a 3-year period.[3]

[1] Units of measure are micrograms per cubic meter of air ($\mu g/m^3$)

[2] TSP = total suspended particles

[3] Annual PM_{10} standard revoked in 2006

Not only can a numerical limit be changed, but also the specific definition of the criteria pollutant. Initially, the NAAQS addressed "total" particulates (PM_{tot}).

In 1987, this was revised to apply to particulates smaller than 10 microns, called PM_{10}. Particles in this size range were considered more likely to cause health effects than larger particles.[17]

Still later (1997), a further refinement of the NAAQS focused on "fine" particulates, defined as particles smaller than 2.5 microns ($PM_{2.5}$). Most recently, concerns have been raised for so-called "condensable" particulate matter (CPM) which are formed (usually solidified) in the atmosphere after being released in vapor phase from a stack or other exhaust source.[18]

There are no formal NAAQS for CPM emissions, but they are considered fine particulates and are often included in $PM_{2.5}$ limitations. All NAAQS targets are routinely re-evaluated and revised as necessary based on up-to-date research.

The net effect of tightening standards and compliance with air emission regulations has been a pattern of steady improvement in ambient concentrations of criteria air pollutants as reflected in Figures 3 through 7.

The dramatic reduction in ambient air SO_2 and lead concentrations were previously shown on Figures 1 and 2.

Figure 3—PM_{10} Air Quality, 1990-2019
(Annual 2nd Maximum 24-Hour Average)
National Trend based on 111 Sites

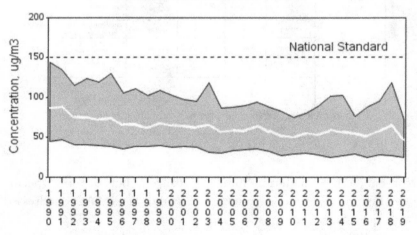

1990 to 2019 : 46% decrease in National Average

Source: U.S. EPA

32

HOLD IT!

Figure 4—PM$_{2.5}$ Air Quality, 2000-2019
(Seasonally-Weighted Annual Average)
National Trend based on 406 Sites

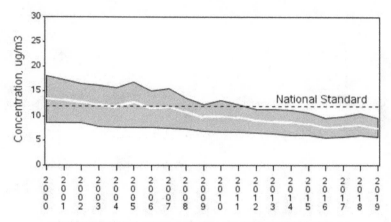

2000 to 2019 : 43% decrease in National Average

Source: U.S. EPA

Figure 5—NO$_2$ Air Quality, 1980-2019
(Annual 98th Percentile of Daily Max 1-Hour Average)
National Trend based on 21 Sites

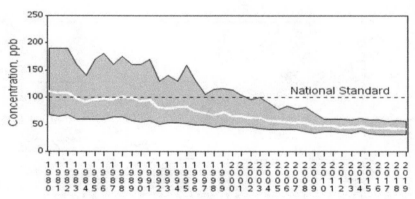

1980 to 2019 : 62% decrease in National Average

Source: U.S. EPA

33

E. Joseph Duckett and Jeffrey L. Pierce

Figure 6—Ozone Air Quality, 1980-2019
(Annual 4th Maximum of Daily Max 8-Hour Average)
National Trend based on 193 Sites

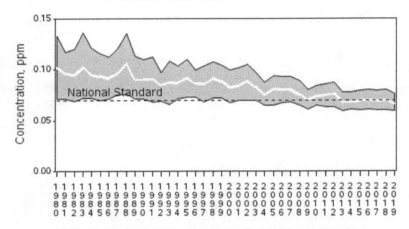

1980 to 2019 : 35% decrease in National Average
Source: U.S. EPA

Figure 7—CO Air Quality, 1980-2019
(Annual 2nd Maximum 8-Hour Average)
National Trend based on 41 Sites

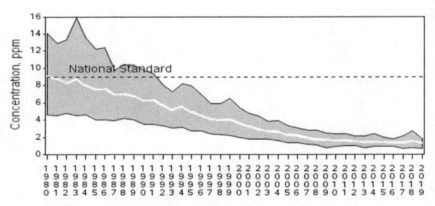

1980 to 2019 : 85% decrease in National Average
Source: U.S. EPA

Figures 1 through 7 reflect national average concentrations over time. Individual areas differ in how closely their air quality has followed the national trends. So-called "hot spots" are local areas which have not improved as much as national averages.

They remain the primary targets for air quality regulations and enforcement in the USA.

Fishable Rivers

When the Clean Water Act was passed in 1972 as an amendment to the 1948 Federal Water Pollution Control Act, one of its stated objectives was to make rivers in the USA fishable and swimmable by 1983.[19]

As has been the case for ambient air pollutants, water quality has improved significantly among most—if not all—rivers.[20]

Source: Midwest Biodiversity Institute; Creative Commons license

On a national average basis, almost all 25 water quality parameters for a variety of measurements have steadily improved over the period 1962 to 2019.

Figure 8 reflects the improvement in water quality among Ohio rivers from 1980 through 2016.[21]

Figure 8—Improved Aquatic Life in Ohio Rivers

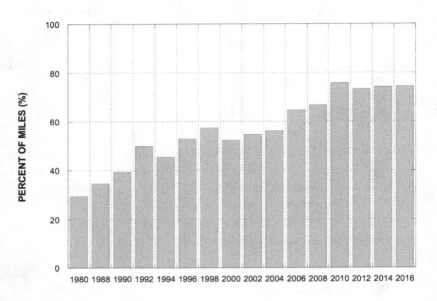

Prime examples of this improved water quality are Pittsburgh, PA's Three Rivers—the Allegheny, Monongahela and Ohio. For many years, these rivers were so polluted that fishing for anything, other than possibly catfish, was discouraged and unproductive. There were simply few fish that could thrive in these rivers.[22]

Reflecting the remarkable transformation of all three rivers, the 2003 National Bassmaster Classic Tournament for professional fisherman was held in Pittsburgh for the first time. A few years later, in 2009, another professional bass fishing contest, the Forest Wood Cup Tournament, was also held in Pittsburgh. In 2020, and throughout the summer months of the past decade, a local outdoors organization hosts weekly public fishing sessions at the "point where the Three Rivers converge." In the fall of 2020, a blue catfish, thought to be regionally extinct for the past 100 years, was caught and released in the Ohio River.[23]

There are at least three major reasons for the water quality improvement of Pittsburgh's rivers.

HOLD IT!

First, increasingly stringent federal, state and local regulations required dischargers to control the quantity and composition of their wastewater.

Secondly, new wastewater treatment facilities have been constructed, both in response to tightening regulations and in sync with increasing recognition of corporate and municipal responsibility for wastewater disposal. Discharges that once had been routed directly into the rivers are now either treated on-site before discharge or piped to large publicly owned treatment works (POTW) for clean-up before discharge.

A third, and not always obvious, factor in improving river water quality, at least in the Pittsburgh area, has been the displacement of industrial facilities to other locations, either within the USA or abroad. Fewer industrial plants translate to fewer wastewater discharges which, in turn, have resulted in improved water quality.

---

[1] Ross, B., and Amter, S., *The Polluters: The Making of Our Chemically Altered Environment*, Oxford University Press, 2010

[2] Lewis, J., *Lead Poisoning: A Historical Perspective*, U.S. EPA Journal, May 1985

[3] Waters, H., *Why Didn't the First Earth Day's Predictions Come True*, Smithsonian Magazine, April 22, 2016

[4] Sebak, R. *Trick or Treatment*, Pittsburgh Magazine, October 1, 2014

[5] *Sulfur Dioxide Trends*, U.S. EPA, June 8, 2020

[6] Clemente, J., *Global Natural Gas Electricity is Gaining on Coal*, Forbes, December 15, 2019

[7] *Cleaned Up Coal and Clean Air: Facts About Air Quality and Coal-Fired Power Plants*, Institute for Energy Research, November 20, 2017

[8] Frazier, R., Study; *Replacing Coal Plants with Natural Gas Cut Pollution, Saved Lives*, State Impact Pennsylvania, January 10, 2020

[9] Pachero, W., *History of Asbestos*, Mesothelioma Center, February 3, 2020

[10] *Workplace Asbestos Regulatory History*, U.S. Center for Asbestos Safety in the Workplace, 2017

[11] Selikoff, I., *Asbestos and Disease*, Elsevier, 1978

[12] *Workplace Asbestos Regulatory History*, Center for Asbestos Safety in the Workplace, 2017

[13] *Asbestosis: Number of Deaths*, National Institute for Occupational Safety and Health, May 2017

[14] Asbestosis, Encyclopedia Britannica, April 2020

[15] *Clean Air Act Requirements and History*, U.S. EPA, January 19, 2017

[16] Table of Historical Sulfur Dioxide National Ambient Air Quality Standards (NAAQS), U.S. Environmental Protection Agency, 2017

[17] *What Are Air Quality Standards for PM?*, U.S. EPA, October 11, 2019

[18] Duckett, E.J., *Condensable Particulates: Troublesome "New" Pollutants*, Iron & Steel Technology, May 2017

[19] *Federal Water Pollution Control Act* (as Amended Nov 27, 2002), U.S. Congress, 33 U.S.C. 1251

[20] *History of the Clean Water Act*, U.S. EPA, June 15, 2020

[21] University of CA-Berkeley, *Clean Water Act Dramatically Cut Pollution in US Waterways*, Science Daily, October 2018

[22] Hopey, D., *30 Years: Pittsburgh's Rivers Show Signs of Life*, Pittsburgh Post-Gazette, October 20, 2013

[23] Hayes, J., *Blue Catfish are Back in Ohio River*, Pittsburgh Post-Gazette, September 17, 2020

CHAPTER 4: EXAGGERATED RISKS

FOR WHATEVER REASONS, many health and/or environmental risks have been exaggerated to be worse than they really are.

Here are some examples...

Drinking Water at Athletic Practices

The 1951 University of San Francisco Dons were one of the top football teams in the country. They had an undefeated 9-0 season and launched some key figures into the National Football League.

For a relatively small private school that dropped football soon after their undefeated season, their roster included three future members of the NFL Hall of Fame (Ollie Matson, Bob St. Clair and Gino Marchetti), at least two other NFL stars (Ed Brown and Dick Stanfel) and a prominent NFL official (Burl Toler). Their coach, Joe Kuharich, later became head coach at Notre Dame and in the NFL. Their publicist, Pete Rozelle, became the Commissioner of the entire NFL during its major expansion.

At the conclusion of their 9-0 season, they were invited to a major bowl game (the Orange Bowl) but declined the invitation on principled grounds. The bowl invitation included a stipulation that they could play

39

the game only if they left their two black players at home (remember, this <u>was</u> 1951). Despite the honor of a major bowl appearance, the team unanimously voted to skip the bowl if it meant playing without two of their players.

Despite all of the remarkable events of that 1951 season, many years later, one of their key players (Bob St. Clair) gave an even more remarkable answer to a reporter's question. He was asked to recall the most memorable aspect of that 1951 season. His answer had nothing to do with winning records or bowl games or even resisting racial discrimination. Instead, he said that his most vivid recollection was that Coach Kuharich never allowed the players to drink any water during practices.[1]

You see, the prevailing wisdom in the 1950's was that drinking water during an athletic practice would cause muscle cramps among the players. This seems absurd today, but it indicates how risks, in this case of drinking water at practices, can be exaggerated and/or incorrect. In fact, the opposite of the prevailing wisdom was true. Today, athletes are encouraged to stay hydrated by drinking water often during practices and games.

Source: Pixabay in iStock;
Creative Commons license

Supersonic Transport (SST)

In the 1970's, the Supersonic Transport (SST) airplanes were heavily promoted as a breakthrough in air transportation. With their first commercial flights in 1976, they were able to fly from London to New York in only 3 hours. The SST planes were expensive and so was the airfare; but they dramatically reduced the travel times between Europe and the USA.[2]

As the SST's were becoming more popular, a potential environmental risk was identified and became the source of some concern.

HOLD IT!

The risk of stratospheric ozone depletion had been the major impetus for regulating chlorofluorocarbons (CFCs) used in refrigerants and other applications. CFCs were acknowledged to be responsible for ozone depletion and were banned worldwide (see Section 5). But, in the case of SST's, the reported risk was not related to CFCs but instead to Nitrogen Oxides (NOx). The release of NOx compounds directly into the stratosphere was thought to be a severe risk to the ozone layer.

It was postulated that the exhaust from the SST's could contain enough NOx to contribute to depletion of the stratospheric ozone layer that protects the earth from excessive ultraviolet radiation.[2] The theory was that NOx from the SST exhaust would act as a catalyst to chemically convert ozone (O_3) to oxygen (O_2).

The SST flew very fast—up to 1,350 miles per hour, twice the speed of sound. The planes also flew higher than other commercial aircraft—up to 71,000 feet above sea level. Their flight paths took them through the lower levels of the stratosphere, the atmospheric layer where ozone predominates.

To address the risk of SST's affecting ozone depletion, several studies were conducted. One particularly alarming 1972 study by a professor at the University of California, Berkeley concluded that within one year (i.e., by 1973) emissions from the SST fleet might halve the concentration of ozone in the stratosphere.[3] He further asserted that this ozone depletion would blind all terrestrial animals, including humans, except animals that lived under water and humans who stayed indoors.

A panel of the National Research Council (NRC) was formed. This NRC Panel issued a report identifying SST's as a major contributor to depletion of the ozone layer. In other words, they confirmed significant risk. They did, however, note that estimates of NOx emissions from SST's were only accurate within an error band multiple of plus-or-minus 10. So the estimates could be 10 times too high or 10 times too low.

The results of the NRC Report were debated. As the debate continued, SST usage and production declined, mostly due to high costs rather than to environmental damage. The bad publicity from the NRC Report didn't help the SST business but, by itself, it also did not kill the SST. The total number of SST's in service never exceeded 14 aircraft. Some of the environmental studies had assumed a fleet of 500 planes.

In 1982 after further research by the NRC and others, the Council had to admit that its 1972 Report was too alarmist and was over-reactive to the ozone depletion effect.[4] The ozone depletion potential of the SST's was estimated to be only about 50% of the earlier estimates, which made it much less of a threat than previously reported. By the time that the NRC revision was issued, the production and use of the SST had dwindled almost to extinction—the victim of costly production costs, ultra-high airfares and low ridership. The use of SST's stopped in 2000. The final straw was a fatal crash at takeoff in France.

Ironically, after more than 20 years of dormancy, United Airlines announced in mid-2021 that they plan to purchase a fleet of 15 new supersonic jetliners. The new planes are being developed by Boom Technology, Inc. Test flights are scheduled for 2021 with full operation by 2029. As with prior SST's, the reported advantage is their flying speed of 1,300 miles per hour. Although reported to be more environmentally "sustainable" than the earlier versions, no mention to date of potential ozone depletion.[5]

There are two added interesting twists concerning stratospheric ozone depletion. The first twist is that ground level ozone is bad (smog), and we do our best to eliminate it through air quality regulations; but stratospheric ozone is good, and we do our best to preserve it.

The second twist is that CFCs were invented by Thomas Midgley, Jr., who also invented TEL, the lead gasoline anti-knocking compound. When he invented these in the 1920s, he thought that he contributed greatly to the well-being of mankind. Instead, some consider him to be the "Dr. Evil" of the environment. We would imagine that if questioned today, Midgley would unapologetically respond, with respect to CFC (which we know as Freon), that his goal was to provide a non-flammable, non-toxic refrigerant to replace the dangerous refrigerants then available to the public. He would claim that Freon saved thousands of lives. One's view of an event or decision is based on who controls the narrative. More about narrative control can be found in Chapter 10.

We also mention that CFCs are typically 5,000 times more potent as a greenhouse gas than CO_2 on a pound per pound basis. The banning of CFCs would have the collateral benefit of reducing greenhouse gas emissions, assuming that the substitute refrigerants have less greenhouse gas potential.

Sometimes, collateral benefits from an environment improvement effort have unexpected collateral environmental benefits. However, this would only be true if the CFC replacement was also not a global warming compound. More on that in Chapter 5.

Alar

In February 1989, the television program "60 Minutes" reported alarming health research findings concerning the chemical Daminozide, trademarked Alar. Alar was an agricultural chemical applied to apple trees over the years 1963-1989 to prevent pre-harvest ripening and increase the storage life of the apples.[6]

The 60 Minutes Report was introduced with a title slide of apples superimposed with skulls and pictures of children in cancer wards. The segment reported that tumors had developed in laboratory mice after receiving doses of Alar and that mathematical modeling indicated cancer risk from consuming

Source: Freepngimg.com; Creative Commons license

apples or apple products from Alar-treated trees. The TV report also noted that Alar residue had been found in some brands of apple juice. The implication was that drinking apple juice could lead to tumor formation in humans, especially children.

There was however more—much more—to the Alar story than what 60 Minutes presented.

Concerns about Alar had originated in 1973, with studies by Bela Toth of the Eppley Institute for Cancer Research in Omaha.

He reported that 1,1-unsymetrical dimethylhydrazine (UDMH) was responsible for the appearance of tumors in laboratory mice. UDMH was a minor contaminant and digestive conversion product of Alar.

As is customary in tumor assays, very high doses of the UDMH test chemical were administered. This practice is to deliberately err on the side of conservatism and safety by detecting potential carcinogenic effects that might not be detected at lower doses.

Two downsides of this approach are: a) that the dosages may be far in excess of realistic human exposures; and b) that the high doses may affect lab animals by acute toxicity rather than by carcinogenicity.

To calculate human risks based on laboratory data, U.S. EPA applies mathematical models and exposure assumptions, again to err conservatively on the side of safety. Only tumor formation data from the most sensitive lab species are included in their modeling. Data from inconclusive or no effect tests are ignored.

Importantly for Alar, all chemicals that appear to induce tumors are treated as genotoxins that cause genetic damage leading to tumor formation.

The Toth studies continued. In 1977, Toth reported a high incidence of tumor formation in mice dosed directly with Alar rather than UDMH. Typically, Alar contained about 1% UDMH. It was known that heat, as used in the production of apple sauce and juice, increased the conversion rate of Alar to UDMH to 5%. Also, U.S. EPA had estimated that humans convert approximately 1% of ingested Alar into UDMH.[7]

In 1980, seven years after the initial Toth results, U.S. EPA commissioned a special review of Alar. After almost 5 years of review, a panel of academic experts concluded that the Toth data were scientifically flawed and were inadequate to provide a basis for banning Alar.

So, in 1986, U.S. EPA allowed continued use of Alar but required additional testing. None of this testing showed an increase in cancer incidence due to Alar or UDMH, even at high doses. The test dose levels were 35,000 times the highest estimate of daily intake of UDMH by preschoolers, the highest risk population. When the dosing of UDMH was increased above accepted toxicity levels, 80% of the male test mice died from <u>toxicity</u>, but without having tumors.

Despite EPA's 1986 conclusion that there was insufficient evidence to ban the use of Alar, several apple product manufacturers (juice; applesauce; baby food) and some major grocery chains had already voluntarily stopped accepting Alar-treated apple products. This was based on the same public concerns that gave rise to the EPA's special review, rather than on a regulatory ban. Many orchard managers had also voluntarily stopped application of Alar.

Revisiting the Toth data in 1989, a British review of that data reached the same conclusion as the U.S. EPA review. The major recognized weakness in the Toth data was that the laboratory dosages were so high that acute toxicity—rather than carcinogenicity—could have produced the observed tumors.

Despite such conclusions about Alar, U.S. EPA announced in February 1989 (the same month as the 60 Minutes Report) that Alar use would be banned after July 1990.[8]

Not content to rely on U.S. EPA's ban, the Natural Resources Defense Council (NRDC) decided to aggressively elevate public fears about the potential cancer risk of Alar.

Based on the discredited Toth data, NRDC performed its own mathematical modeling to calculate a higher cancer risk than U.S. EPA. The NRDC model assumed that both UDMH and Alar were genotoxic (causing genetic mutations) even though both had been shown <u>not</u> to be genotoxins.

The previously noted British review had considered the lack of genotoxicity coupled with low human exposures to declare Alar safe. Around the same time, a study by the California Department of Food and Agriculture (DFA) conducted tests beyond the Toth data. They concluded that Alar restrictions should be limited to apple juice and apple sauce, not to raw apples or other apple products.[9]

Again, not content to rely on even-handed scientific analyses and despite the fact that Alar had not been identified as the cause of a single childhood cancer, the NRDC promoted an alarmist message painting Alar-treated apples as a major cancer risk for children.

NRDC hired a public relations firm to both publicize their own conclusions and to plant the cancer risk story at 60 Minutes. The 60 Minutes story was completely one-sided with, as one author put it: "utter disregard for objective reporting."[8]

At the end of this Alar episode, many school systems removed apples and apple products from their lunches.

Apple sales plummeted.

All sales of Alar ceased by December 1989.

Applying the conservative California DFA modeling estimates of probable risk from Alar, the population of the USA would have to increase 1000-fold before even one Alar-caused tumor would be expected. A later study showed that the Alar dose to which the lab mice were exposed was equivalent to the human equivalent of drinking 5,000 gallons/day of apple juice.

Obviously, an exaggerated threat but the negative publicity was enough to scuttle the use of this safe, effective agricultural tool.

A later calculation showed that a person consuming 4.5 pounds of applesauce per day would, in one year, ingest an amount of UDMH equivalent to smoking two filtered cigarettes in a lifetime.

Exaggerated risk...indeed.

Polio and Ice Cream

In the 1950's, polio was a much feared—and very real—threat to human health.[10]

The disease primarily struck children, rendering them paralyzed, and often needing mechanical devices called "iron lungs" just to breathe.

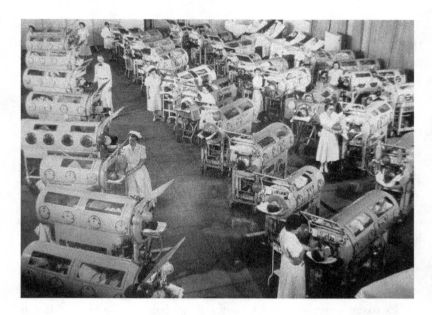

Source: Brewininate 7/8/18; Creative Commons license

Until Jonas Salk discovered and tested an effective vaccine in 1954, polio gripped the country. As often happens in the midst of such panic, unscientific theories about the disease began to circulate.

One such theory was that polio was transmitted by ice cream. Today, this theory seems outrageous and has been thoroughly debunked. But, in the midst of the polio fear crisis, even unlikely theories were pursued.

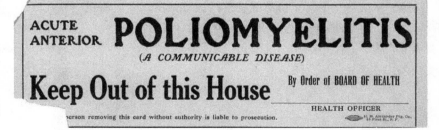

Source: Wikimedia Commons; Creative Commons license

The hypothetical link between polio and ice cream spun out of the fact that polio affected mostly children and seemed to be transmitted primarily in warm summer months.

Children like ice cream in the summer so, as the conjecture went, maybe that's how polio spreads.[11]

As with other baseless scares, it was much easier to arouse fears than to debunk the theory. Research into the possible association was conducted. After years of testing, the theory was dismissed, although some of the fear lingered longer because it is almost never possible to "prove" a negative association.

Fears almost always subside slowly.

This example illustrates that a correlation between exposure and health does not necessarily prove causation.

New Ice Age

As discussed extensively in Chapter 9, global climate change is a hot topic (literally!) in recent years.

The concern is that warming temperatures throughout the world will lead to a variety of environmental and public health problems.

In the early to mid-1970's, average worldwide temperatures actually dropped for several years.[12]

Over the period 1964-1975, average global temperature decreased in 9 of 13 years.[13]

Although not as prevalent as concerns for global warming today, there were serious concerns for global cooling.

As noted in the Chapter 1 of this book, an article in the April 28, 1975 issue of Newsweek Magazine ran under the headline *The Cooling World* and alarmingly stated that the cooling would have "terrible consequences."

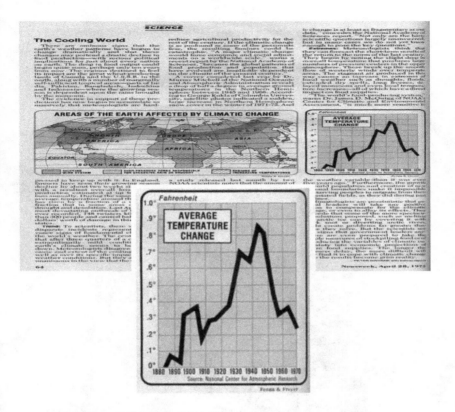

Source: Wikimedia 9/14/21; Creative Commons license

One concern among environmental and climate scientists in the 1970s was that increased atmospheric build-up of pollutants (especially particulates) was shielding sunlight from the earth, thereby leading to cooler temperatures.

In retrospect, this theory of shading from particulate air pollution may have had some merit. As noted in Chapter 3, atmospheric particulate levels have dropped significantly since the 1970's. One of the most peculiar suggested approaches to reducing global warming is to artificially increase the airborne particulate concentrations to shield/deflect incoming sunlight.[14] No responsible governmental entity has bought onto this approach. But, here we seem to go again. The environmental goal of reducing airborne particulates seems to have had the unintended consequence of contributing to another environmental "crisis."

Whatever the causes of "global cooling," if in fact it was occurring, it is now obvious that the cooling trend has been reversed, and the next "Ice Age" seems very distant. Nevertheless, more recently, Discover magazine ran a cover story on "A New Ice Age" in 2002 and reported that there may be a new risk of Northern Hemisphere cooling but as a <u>result</u> of global warming. They referenced scientists who speculated that arctic ice melting, a consequence of global <u>warming</u>, could release enough fresh water into the oceans to trigger Northern Hemispheric <u>cooling</u> by subverting the northern penetration of Gulf Stream waters.

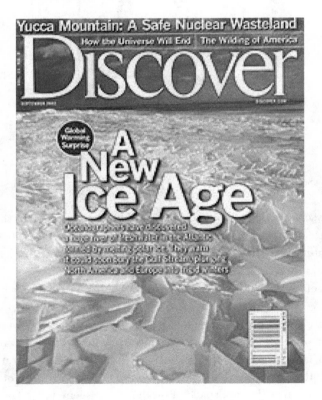

Source: Quora 9/2017; Creative Commons license

Lead in Drinking Water

As noted in Chapter 3, ambient levels of atmospheric lead have been dramatically reduced, largely through the abandonment of using leaded gasoline.

Lead, however, is still considered a serious health hazard, especially for children. The effects of lead exposure on childhood learning and development have been well documented and such effects are not exaggerated.[15] As noted in Ben Franklin's quotation from Chapter 1, recognition of lead as an occupational "poison" affecting printers, painters and plumbers dates back at least to the early 1800's. Aside from occupational exposures, most of the more recent focus has been on children under 6.

Source: R Street Institute 2016; Creative Commons license

Exaggeration, in the case of childhood lead exposures, applies not to the environmental health risk of lead, but to the sources of lead ingestion.

Flint, MI became notorious in recent years because of a switch in drinking water supplies in 2014. The source of their water temporarily changed from Lake Huron (as treated and supplied to Flint by the City of Detroit) to the Flint River (as treated directly by the City of Flint itself).

Chemically, the Flint River water was more corrosive than the Huron/Detroit water. As a result, the more corrosive water leached metals from leaded supply pipes in Flint. Lead levels in Flint water rose notably and created a remarkable blizzard of alarming stories of poisoned children and environmental distress.

Subsequent testing of drinking water from many other cities in the USA and Canada, including other Michigan cities, revealed higher than expected lead levels in almost all cities with aging water supply piping.[16,17]

Lead pipes and leaded solders for water supply plumbing began being phased out around 1930, and were effectively banned by the 1986 amendments of the Safe Drinking Water Act.[18]

Two factors reveal the exaggeration of childhood lead exposures from drinking water.

First, there has been a 90% drop in measured childhood lead levels nationwide from 1976–2004.

If anything, this does not suggest an increasing risk.

Secondly and more importantly, despite the widely publicized lead contamination problem in Flint (and other cities), study after study has shown that most lead ingestion among children is from deteriorated lead paint and lead paint dusts, not drinking water.[19,20]

Intake studies among children have revealed that up to 90% of lead ingestion is from paint rather than other sources (including drinking water).[21]

Among eight potential sources of lead from a 2018 study in Ohio, water was ranked eighth below such other sources as cosmetics, food and toys.[22]

HOLD IT!

Figure 9

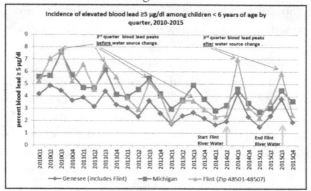

Source: MDPI, International Journal of Environmental Research and Public Health, 3/18/2016, Creative Commons license

Figure 10

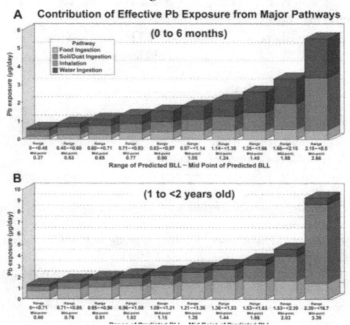

Source: U.S. EPA

E. Joseph Duckett and Jeffrey L. Pierce

Lead-based paints for use in houses were banned in 1978. But prior to that, many homes contained lead-based paint. Once painted, the lead was still present in these homes even if over-painted with non-lead paints. As paint deteriorates over time, the leaded paint flakes and dust can be a source of lead ingestion by young children.

Children are at a rapidly developing stage of life and are very susceptible to the effects of lead ingestion. Due to an eating phenomenon termed "pica," some children actually seek and ingest flakes of paint, reportedly due to a calcium deficiency. In effect, as the theory goes, children ingest paint because it may contain calcium. Because the paint often also contains lead, the children ingest lead as an unintended consequence of trying to increase their calcium supply.

So, in the case of childhood lead poisoning, the exaggeration is not that this condition is falsely alarming but that it's attribution to public drinking water is overstated. Children consume lots of fluids (milk, juices, even bottled drinking water) that are unrelated to their local tap water. By focusing too much alarm on tap water sources, attention may be diverted from the real major sources.

To seriously address childhood lead levels, the focus should be on old lead-painted walls more than on old lead pipes.

---

[1] Clark, K.S., *Undefeated, Untied and Uninvited*, Griffin Publishing Group, Irvine, CA, 2002

[2] Concorde, Encyclopedia Britannica, May 8, 2020

[3] Sullivan, W., *Science Panel Says SST's Could Endanger Earth*, New York Times, November 5, 1972

[4] *Scientists Moderate Estimates of Ozone Depletion*, New York Times, April 1, 1982

[5] Grossman, M. and Sider, A., *United Plans to Buy 15 Supersonic Planes*, Wall Street Journal, June 3, 2021

[6] Milloy, Su, *Background on the ALAR Controversy*, West Lake Solutions, 1998

[7] Shaw, D., *Alar Panic Shows Power of Media to Trigger Fear*, Los Angeles Times, September 12, 1994

[8] Rosen, J.D., *Much Ado About Alar*, Issues in Science and

Technology, Fall 1990

[9] California Department of Food and Agriculture, *Analysis of Natural Resources Defense Council Report—Intolerable Risk: Pesticides in Our Children's Food*, Sacramento, CA, May 25, 1989

[10] *Polio: A 20th Century Epidemic*, Science Museum (London), Oct. 15, 2018

[11] Dupuy, T., *Vaccinations and Ice Cream Scare*, Column, July 2011

[12] Global Temperature, USA NASA Goddard Institute for Space Studies, 2020

[13] Lindsey, R. and Dohlman, L., *Climate Change: Global Temperature*, National Oceanic and Atmospheric Administration (NOAA), August 14, 2020

[14] Wagner, G., *The Fast, Cheap and Scary Way to Cool the Planet*, Bloomberg, June 3, 2020

[15] Rabin, R., *The Lead Industry and Lead Water Pipes : A Modest Campaign*, America Journal of Public Health, September 2008

[16] Nieves, A., *Cities Across the US Are At Risk of Being Exposed to Elevated Levels of Lead in Their Water*, WXYZ National News, January 3, 2020

[17] Leggett, C., *Tap Water in Some Canadian Cities Has As Much Lead Contamination as Flint, Michigan*, NARCITY, October 2019

[18] Hings, F., *Use of Lead Free Pipes, Fixtures, Solder and Flux for Drinking Water*, US EPA, January 27, 2020

[19] Advisory Committee on Childhood Lead Poisoning Prevention, *Preventing Lead Exposure in Young Children*, October 2004

[20] *Childhood Lead Poisoning Prevention*, U.S. Centers for Disease Control, 2019

[21] *5 Surprising Sources of Lead Exposure*, WebMD, October 2019

[22] *About Lead*, Ohio Department of Health, November 2018

CHAPTER 5: UNINTENDED CONSEQUENCES

EVEN WITH THE best intentions, actions to resolve one environmental or public health problem sometimes trigger another, often new, problem.

Some have been mentioned above and additional examples are provided in this chapter.

Saturated Fats

In the 1980's, the US Food and Drug Administration (FDA) issued a recommendation against dietary saturated fats.

They instead recommended unsaturated "trans" fats as the alternative.[1]

Flash forward to 2015, and the same FDA ordered removal of trans fats from all foods.

Turns out that the alternative was worse than the original saturated fats "problem."[2]

CAFE Standards

In the 1975 Energy Policy & Conservation Act, Congress authorized the U.S. EPA to set fuel economy standards for vehicles in the USA. These were called the Corporate Average Fuel Economy (CAFE) standards.

They set the fleet average miles-per-gallon (mpg) minimum requirements for all vehicles to be sold in the USA.

Setting these standards became controversial. The intention was to constrain gasoline consumption and hence, reduce dependence on foreign oil supplies while also reducing vehicular emissions. Rather than setting a single standard for all vehicles, U.S. EPA set two different standards applicable to most vehicles.

The CAFE standard for passenger cars was set at 27.5 mpg. A different, more lenient, standard was set to apply to "light trucks." When these initial distinctions were made, only about 20% of all passenger vehicles were considered "light" trucks, so the tighter standard (27.5 mpg) was intended to apply to most vehicles.[3]

But under the definition of "light" trucks, a new type of passenger vehicle was included. This new passenger vehicle was the Sport Utility Vehicle (SUV). Based on the average weight of SUV's, they were under the less restrictive standard, 20.7 mpg. This meant that SUV's only had to be 75% as fuel efficient as passenger cars to comply with the CAFE rules. To this day, when Jeff wants to use their 2005 Mercedes 350 ML, he tells his wife he is "taking the truck."

As of 2019, a full 70% of all new passenger vehicles in the USA are SUV's. This is a more than 300% increase from pre-CAFE levels, at least in part due to the relaxed mpg standards under CAFE. So, in a reversal of the original regulatory targeting, the more restrictive standard ended up applying to a much smaller portion of the national vehicle fleet.

In the words of one author:

> *...the steady increase in light-truck sales, largely due to lower fuel economy standards for trucks and SUV's, actually drove down fleet-wide fuel efficiency...*[4]

MTBE

The struggle to control environmental lead exposures has extended into several regulatory programs and control strategies. As noted in Chapter 3, the removal of tetraethyl lead (TEL) from gasoline was a major contributor to reducing ambient lead concentrations—even if the removal was largely due to concerns for fouling catalytic convertors.

For many gasoline blends, the substitute for TEL to reduce engine knocking was Methyl Tertiary Butyl Ether (MTBE). By 1979, it was commonly used as the preferred TEL replacement.

MTBE can be readily blended with gasoline. It has low volatility and is inexpensive. These were all advantages of using it to replace TEL. Another attribute that seemed innocuous at first but later became disadvantageous was that MTBE is water soluble.

In 1992, the Clean Air Act included a requirement that oxygenated fuels had to be used seasonally in areas with CO and/or ozone exceedances. In 1995, U.S. EPA required the use of reformulated gasoline year-round. U.S. EPA did not specifically require the use of MTBE, but almost 90% of all reformulated gasoline did contain MTBE.

First observed in Santa Monica, California, some underground MTBE storage tanks were found to be leaking into groundwater. The water-soluble MTBE contaminated well water supplies. To date, MTBE leakage problems have been severe enough that the additive is now banned in 25 states.[5]

The primary replacement for MTBE has been ethanol. So far, ethanol has been considered to be environmentally friendly. However, there are competing opinions as to whether over the entire production-to-use lifecycle of ethanol, the carbon intensity of ethanol exceeds, is equal to, or is lower than gasoline; and whether the net production-to-use lifecycle emissions of specific criteria pollutants are lower or higher for gasoline or ethanol.

NOx Control and SO₃

As the ambient standards for ozone have been tightened, so have the restrictions on the two major precursors of ozone—volatile organic compounds (VOCs) and oxides of nitrogen (NOx).[6]

To meet the tightened standards for NOx emissions, many combustion facilities (power plants, boilers and large gas turbines) have been equipped with selective catalytic reduction (SCR) which successfully reduces NOx emissions.

As SCR's were applied to coal-fired power plants or boilers, it was observed that, in addition to reducing NOx, the control system also converted gaseous sulfur dioxide (SO_2) to particulate sulfates. These sulfates are very difficult to control because they are typically sub-micron sized, condensable particulates. These are not easily captured in conventional particulate emission control equipment, such as baghouses or electrostatic precipitators. When sulfates are released from an emission stack, they tend to form a visible blue plume.[7]

Figure 11

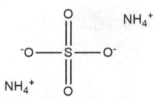

Ammonium sulfate

Source: Wikipedia 2013; Creative Commons license

Added to the difficulty of capturing sulfate particulates is the problem of atmospheric formation of sulfuric acid when the sulfates react with water. A solution to NOx emissions has resulted in the unintended formation and release of particulate sulfates, which in turn converts to atmospheric sulfuric acid.

Recycling and Contaminants

Reuse and recycling of municipal wastes has been promoted for at least the past 50 years. Adopting practices to recover materials of value from waste streams can reduce the need for disposal, relieve the demand for virgin raw materials and often conserve energy.

Source: SVG Source; Creative Commons license

For at least one waste material— glass bottles—governmental pressures to increase recycling rates have actually reduced the quality and restricted industrial consumption of the recycled materials. Waste glass for use in glassmaking is called "cullet."

Glass container recycling is complicated on many fronts. Firstly, glass containers come in three major colors: flint (clear), green and amber. Flint glass is predominant in the USA, accounting for more than 60% of all container glass. The chemical colorants used to produce green and amber glass are unwanted contaminants of the cullet to be used in flint glass manufacture.[8]

Secondly, glassmaking operations are very sensitive to solid refractory pieces that don't completely melt in a glass furnace. If these particles survive the furnace and get incorporated into a glass bottle, they are called "inclusions." Such inclusions obviously affect the appearance of the bottle. Less obvious is the fact that inclusions can create a serious safety hazard if they result in a stress concentration, causing a bottle to burst under pressure.[9]

To remedy the color issue, most successful glass container recycling systems sort the bottles by color before being sent to a glass manufacturer. Often, the color separation is performed at an intermediate processing facility between local collection centers and the glass plants.[10]

For obvious reasons (worker safety is a big issue), broken glass cullet is more difficult to color sort than whole intact bottles. Color sorting broken glass requires more hand sorting care. Methods to adjust glass color within glassmaking furnaces have been developed; but, again, they add complication to the overall waste glass cullet recycling process.

Without color separation, the potential for reuse of waste glass cullet is essentially limited to production of colored bottles (green or amber) rather than flint. The tolerance for colorants in flint glass is very limited. Techniques for de-colorizing colored cullet have been developed but such additional processing increases the costs of recycling.[11]

To avoid refractory contaminants in waste glass cullet, hand-sorting removal of ceramics (e.g. coffee cups) and other materials (stones, rocks, etc.) can be employed—once again adding to the processing costs.

Another approach to reducing refractory-based glass inclusions is to increase glass furnace temperatures and/or lengthen the melting cycles. Either of these approaches adds to the costs of glassmaking and can lessen any energy savings from cullet use, thereby reducing the recyclability of the cullet.

Given the importance of cullet quality (color; refractories), successful reuse of glass bottles recovered from municipal waste requires careful quality control.

Unfortunately, most governmental rules requiring bottle recycling emphasize the quantity of recycling rather than quality. As noted above, the need for better quality has spawned an "industry" of intermediate waste processors. In turn, the added processing costs have economically discouraged use of the cullet.

An original purpose of municipal waste recycling—including glass bottle recycling—was that the recovered materials would have sufficient value to warrant the extra costs of collection and processing. In other words, there was supposed to be an economic, as well as environmental, justification for recycling wastes that were otherwise destined for disposal.

This promise was abandoned years ago when municipalities began making recycling mandatory. Typically, municipalities now pay extra fees, beyond disposal costs, for recycling wastes.

HOLD IT!

Although many municipalities post guidelines for household separation of recyclables, mis-separation is inevitable in a public collection program. Some jurisdictions have even assigned inspectors to enforce waste separation regulations.[12,13]

This, of course, further increases the costs of recycling. Together with the costs of processing recyclables, the total costs of recycling have been reported to be as much as 25% higher than the costs of conventional collections and disposal.[14] Since recycling enforcement cannot be perfect, the misclassification of recyclables and contamination with non-recyclable wastes has increased as the push for increased recycling quotas has proceeded.

In addition to quality deterioration, the drive to increase quantities of recycled wastes has also resulted in localized supply-demand imbalances where there is no local glass plant to absorb locally-collected cullet.

In the case of waste glass cullet, the costs of transporting a dense material like cullet from a recycling facility through a cullet processor to a distant glassmaking plant can be a deal breaking barrier to reuse of recovered cullet.[15]

In cases where regulatory quotas, rather than market prices, were driving recycling rates, the added expenses of processing and shipping have to be absorbed into the recycling programs, either by the collecting municipalities or by the cullet users.

For some recyclables, such as paper and plastics, international exports offered a solution to domestic market imbalances. The USA export market for cullet, however, was never robust, owing in large part to high shipping costs. Cullet is a relatively low value, high density material that is costly to transport domestically, even more so internationally.

The combined effects of reduced quality, expensive processing/shipping and the absence of an export market have resulted in some municipalities eliminating bottle recycling from mixed household wastes. So, the drive to <u>increase</u> collection of recyclables with little regard for markets or quality has now led to <u>reduced</u> recycling of glass bottles.

An ironic twist to the fall-off in glass recycling is related to the COVID-19 Pandemic. With the effective vaccine having been found, its distribution requires billions of clear (flint) glass vials.[16] Producing such a large quantity of glass would require a considerable jump in high quality cullet supply. As always, market fluctuations are somewhat unpredictable—even for materials recovered from municipal wastes.

But, here is the point.

Why recycle glass bottles?

Landfilling glass bottles is not a problem.

Glass is environmentally inert, so there is no problem with disposing of them in landfill, and there has never been a landfill capacity problem.

If there is no, or only a limited, market for recycled glass bottles, why try to recycle them? When Jeff and Joe were kids, all soda and beer was sold in glass bottles. You returned them to the seller if you wanted cash back, and the seller would return the bottles for reuse. You could search the neighborhood for discarded bottles and return them to the local store for two cents or five cents to buy penny candy. Jeff's first experience with rejection on the return of glass bottles for money was when he tried to return a case of empty Heineken bottles to a beer distributor in Pittsburgh in the late-1970's when the guy asked—what do you expect me to do with these, send them back to Holland?

CFCs/HFCs

The Montreal Protocol on Substances that Deplete the Ozone Layer (Montreal Protocol) was signed in 1987. It primarily sought elimination of the use of chlorofluorocarbons (CFCs), a class of very stable chemicals used as refrigerants and aerosol propellants. CFCs can react with sunlight and breakdown stratospheric ozone.

To date, the use of CFCs has been almost completely eliminated.[17]

Figure 12

Ozone Layer Depletion by CFCs

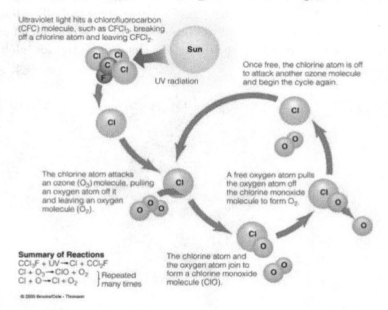

Source: Fab Pretty, 4/18/2013; Creative Commons license

Some have suggested that the success of the Montreal Protocol in eliminating CFCs could point to a successful approach to reducing greenhouse gas (GHG) emissions. For several reasons, the CFC experience is not easily repeatable for controlling other pollutants.[18]

First, at the time of the Montreal Agreement, CFCs were already out of patent and considered old technology. So, there were reasons to transfer from CFCs to other refrigerants beyond concerns for ozone depletion.

Secondly, there were a limited number of CFC producers, thus requiring a much smaller number of signatory agreement countries, than for other air pollutants. For example, only four countries accounted for almost two-thirds of all global CFC production at the time of the Montreal Protocol.

Thirdly, the large number of CFC user countries had little choice but to change from CFCs once they were no longer being produced.

Despite the success of the 1987 Montreal Protocol, the global ozone layer is only expected to return to 1980 levels sometime between 2045 and 2060, indicating the long-time lag for global adjustments.

Source: Aircetera 6/2/06; Creative Commons license

In most applications, hydrofluorocarbons (HFCs) have replaced CFCs. The corresponding dramatic increase in HFC production has resulted in an apparently unanticipated consequence. HFCs don't attack ozone, but like CFCs, they are greenhouse gases. Both CFCs and HFCs are thousands of times more potent (pound for pound) than CO_2, as a greenhouse gas, based on their 100-year warming potential.[17] This potential considers both the infrared absorption of the gas and their expected atmospheric lifetime. HFCs have similar, and sometimes higher, global warming potential than CFCs. Control of global ozone depletion has spawned, or at least encouraged, a threat to control of global climate change. As a minimum, resolution of the ozone depletion problem turned a blind eye toward another developing environmental concern.

As noted in Chapter 4, CFCs were invented by Thomas Midgley, Jr., an employee of General Motors' Frigidaire Division. He was seeking a non-toxic, non-flammable substitute for the refrigerants that were in use in the 1920s (e.g., ammonia, propane and sulfur dioxide). He named the gas that he invented Freon. He had previously invented the lead-based gasoline additive TEL. As explained above, TEL prevented reciprocating engine knocking. Both inventions were wildly successful products, and they served their intended purpose. In his time, he was lauded as one of the greatest inventors of his age. His legacy has been greatly tarnished by the environmental impact of these products. CFCs and TEL are now routinely found on lists of the "Worst Inventions of All Time."

U.S. EPA Cleanup Spill

In August 2015, the U.S. EPA was in the midst of "cleaning up" contamination from the Gold King Mine in Colorado. The mine was abandoned at the time of the U.S. EPA project.

U.S. EPA had determined that the wastewater impoundment at the mine contained lead, copper, arsenic and cadmium toxic metals and was acidic. Most of the toxic metals were believed to be in sludge which had settled to the bottom of the impoundment.

Source: Wikimedia by Riverhugger,8/6/2015; Creative Commons license

During the cleanup, things went terribly wrong, and the impoundment dam was breached, releasing the contaminated water and sludge into the Animas River.[19] This river is a tributary to the Colorado River and passes through New Mexico, Arizona and Utah.

In an ironic reversal of roles, U.S. EPA itself was the responsible "polluter" in this event. U.S. EPA, in a manner often ascribed to industrial dischargers, adopted what the New Mexico Environmental Secretary called a "cavalier and irresponsible" attitude in their response to the spill.

The Navajo Nation even sued U.S. EPA for damages to their water supply.

U.S. EPA ended up warning people to stay out of the Animas River and to keep domestic animals from drinking its water.

Many of U.S. EPA's responses to the spill reflected the same type of defensiveness that they, U.S. EPA, often accuse industries or municipal dischargers of displaying. U.S. EPA delayed notification of local officials by more than 24 hours. Even then, they reported a spill volume that was only about one-third of the actual release.

In fact, the initial notification came from the Southern Ute Tribe—not from U.S. EPA directly. Neither U.S. EPA nor their contractor had a contingency plan to activate in the event of a spill. U.S. EPA assigned only two staff members to work with local Tribes to analyze causes and take corrective actions.

An Interior Department investigation concluded that U.S. EPA rushed the cleanup work, minimized engineering/planning to protect the dam, and did not adequately check water levels in the impoundment.[20]

The eventual cleanup costs approached $4.5 million but U.S. EPA, on the basis of sovereign immunity, rejected an additional $1.2 billion in claims from farmers, states and others affected by the spill.[21]

In the aftermath of this spill, U.S. EPA suspended similar site investigations and cleanup at 10 mining complexes in four states.

Botox

Not all unexpected consequences produce negative results. The Botox story illustrates how an acknowledged powerful toxin can be repurposed as a beneficial medicine and cosmetic.

Clostridium botulinum is an anaerobic bacterium. It has been, and still is, recognized as one of the most dangerous sources of life-threatening food poisoning. Canned foods, especially home canned, that are low in acidity (e.g., fruits and vegetables) and prepared under low oxygen conditions can allow botulism toxin to be produced. Botulism toxin is a potent neurotoxin and muscle relaxant. As a food poison, it can cause respiratory and muscular paralysis.

Unlike other bacterial food contaminants, like Salmonella, Escherichia (E) coli or Staphylococcus aureus, the food poisoning effect of C. botulinum results from the toxin produced by the bacteria, rather than from the organisms themselves. Consequently, foods contaminated with botulism toxin cannot be eaten safely unless thoroughly cooked to temperatures above 185° F (at least 20° F above normally accepted safe temperature for cooking or reheating food). This makes botulism an especially troublesome food poison.

How ironic then that a protein made from this same Botulism toxin, now reincarnated as Botox, has become extremely popular as a medicine and cosmetic. It is injected to reduce the wrinkles of aging skin and to protect against migraine headaches and neck spasms.[22]

Health risks often stem from how and where a substance is used as much as what it is. A characteristic, in this case neurotoxicity, can swing from risk to benefit based on its application and exposure.

---

[1] *Trans Fat*, U.S. Food & drug Administration (FDA), May 18, 2016

[2] *Saturated Fat vs. Trans Fat: Which is Worse?*, National Bio Health, June 23, 2015

[3] *Corporate Average Fuel Economy (CAFE)*, National Highway Traffic Safety Administration, September 29, 2020

[4] Lubetsky, J., *History of Fuel Economy*, PEW Environment Group, April 2011

[5] *MTBE Factsheet*, U.S. Centers for Disease Control and Prevention (CDC), April 7, 2017

[6] Duckett, E.J., *Advanced NOx Controls for Industrial Sources*, Proceeding of Air & Waste Management Association Annual Conference and Exposition, June 2001

[7] Buecker, B., *Emissions Control: SCR Design and Operation*, Power Engineering, August 1, 2002

[8] Duckett, E.J., *Glass Recovery and Reuse*, National Center for Resource Recovery Bulletin, Fall 1978

[9] Tooley, F., *The Handbook of Glass Manufacture*, 3rd Edition, Ashlee Publishing Co., 1984

[10] Duckett, E.J., *The Influence of Color Mixture on the Use of Cullet Recovered from Municipal Solid Waste*, Conservation & Recycling, June 1979

[11] Duckett, E.J., *Cullet from Municipal Waste*, Ceramic Industry, March 1979

[12] List, M., *Trash Police?*, Providence Journal, February 8, 2019

[13] Sridhar, P., *Proposed $50 Penalty for Improper Disposal of Dirty Diapers*, KENS News, February 13, 2018

[14] Rue, D., *Cullet Supply Issues and Technologies*, Glass Manufacturing Industry Council, October 2018

[15] Jacoby, M., *Why Glass Recycling in the US is Broken*, Chemical & Engineering News, February 11, 2019

[16] Peters, A., *The COVID-19 Vaccine will require billions of glass vials—but lockdowns are cratering the supply of recycled glass*, Fast Company, May 2020

[17] Fisher, J. and Wilson, S., *Explainer: Hydro Fluorocarbons Saved the Ozone Layer, So Why are we Banning Them?*, The Conversation, November 1, 2017

[18] Rae, I., *Saving the Ozone Layer: Why the Montreal Protocol Worked*, The Conversation, September 9, 2012

[19] Schlanger, Z., *EPA Causes Massive Spill of Mining Waste Water in Colorado; Turns Animas River Bright Orange*, Newsweek, August 7, 2015

[20] Plautz, J., *Government Report Faults EPA in Colorado Mine*

Spill, The Atlantic, October 22, 2015

[21] *EPA Says it Won't Repay Claims for Spill that Caused Yellow Rivers*, CBS News, January, 13, 2017

[22] Mayo Clinic Staff, *Botox Injections*, Mayo Clinic Website, 2021

E. Joseph Duckett and Jeffrey L. Pierce

CHAPTER 6: REGULATORY OVERREACH

Waters of the USA

WHEN IT WAS first written into the Clean Water Act of 1972, the phrase "waters of the United States," to describe waters to which the law was applicable, seemed straightforward and non-controversial. In the ensuing years, as the law was converted into regulations, and subjected to multiple court challenges, the definition of waters of the United States was much debated and became very controversial.

The principal focus of debate has been the extent of the reach of the waters of the United States, more specifically, the division of regulatory jurisdictions between federal and state authorities. Clearly large navigable rivers, lakes and major tributaries were covered by the Act, but non-navigable waters, to many parties did not appear to be waters of the United States. A succession of regulatory actions and court decisions, at both the federal and state levels, ultimately defined the type and size of waterways governed by the U.S. EPA rules.

In 2015, the escalating expansion of defining "waters of the United States," resulted in an U.S. EPA rule that extended federal jurisdiction into previously state-regulated bodies of water.[1] The 2015 rule even included temporary or transient wetlands and small streams that may only flow during rain events. In other words, the definition of waters of the United States was expanded enough to cover areas that might become water channels, but did not necessarily contain any water much of the time.

In April 2020, the U.S. EPA and Army Corps of Engineers jointly announced a repeal of the 2015 rule which they contended had been an overreach.[2] The revised regulation is called the "Navigable Waters Protection Rule." Repeal of the previous regulation came on the heels of two federal district court rulings that had remanded the 2015 rule back to the agencies.

The controversy over navigable waters typifies the confusion that results from the legislative branch's passage of a law, and then the executive branch's implementation of the law, and then the warring parties asking the judicial branch to decide what the legislative branch intended. Isn't the simple solution to ask the legislative branch to clarify what they intended—to clarify the law by amendment? Congress gave Federal agencies broad regulatory authority over pollution of navigable waters. Granting that authority would broadly be constitutional under the U.S. Constitution—polluted waters could interfere with the interstate transfer of goods or could affect another state's beneficial use of water. But what is a navigable water? The agencies, in their chase for control of all point source and non-point sources of water pollutants claimed that streams that can intermittently pass a shallow canoe and all wetlands (pockets of water) are navigable waters. It would not seem to cross the minds of the parties battling in court (the U.S. EPA and the regulated parties) to ask Congress to clarify the definition of navigable water—to clarify where they wanted to have the law apply. And Congress seems to have no interest in revisiting things that they may not have fully understood when they passed the law.

Mercury Hot Spot Emission Rules

Air emissions of mercury are both unusual and controversial. Mercury itself is unusual as the only metal that is liquid at room temperature. Controversy has raged for years over the extent of mercury emission reductions required to achieve safe levels and the regulatory mechanism by which such reductions should be accomplished. The principal options were either across-the-board stack emission reductions or adoption of a cap-and-trade program (discussed in Chapter 10). In many respects, these controversies derive from the unusual nature of mercury as a pollutant.[3]

Unlike most other regulated air pollutants, almost everyone agrees that the risk to human health from the direct inhalation of mercury now present in ambient air is well below any level of concern.[4] Mercury is simply not viewed as an inhalation public health threat.

Figure 13—Simplified Mercury Transport, Conversion and
Bioaccumulation Cycle

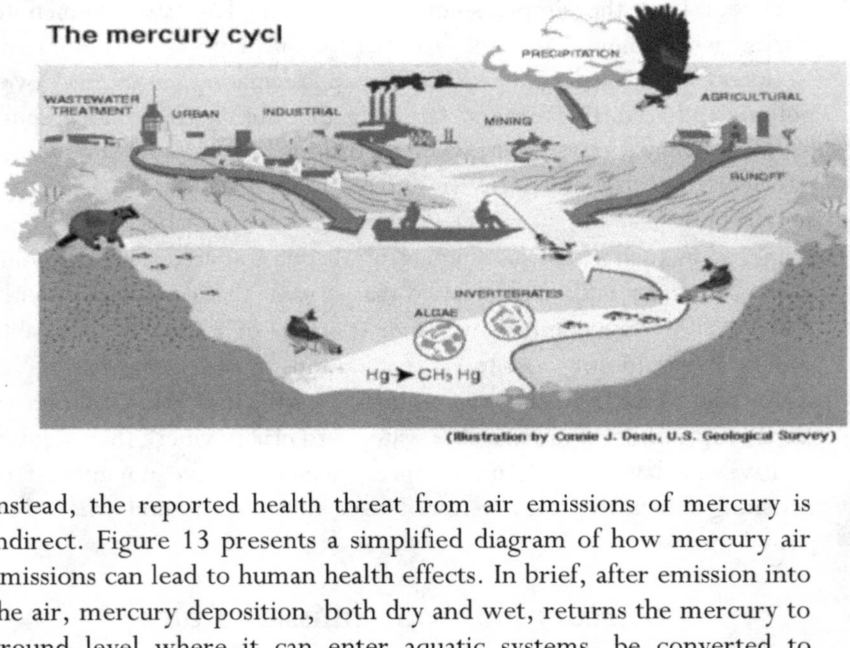

(Illustration by Connie J. Dean, U.S. Geological Survey)

Instead, the reported health threat from air emissions of mercury is indirect. Figure 13 presents a simplified diagram of how mercury air emissions can lead to human health effects. In brief, after emission into the air, mercury deposition, both dry and wet, returns the mercury to ground level where it can enter aquatic systems, be converted to methyl mercury chloride and bioaccumulate through the food chain, resulting in elevated mercury concentrations in fish tissue. The link between mercury air emissions and health effects therefore depends upon a series of factors: emission rates, atmospheric chemistry, deposition rates, entry into waterways, aquatic conversion, bioaccumulation, fish consumption and human health effects. This is a long path influenced by many variables and stretched over time.

As for the health effects of fish consumption, the principal concern is eating fish containing methyl mercury that has bio-accumulated up the food chain.[5] Fish advisories have been issued for many bodies of water across the USA, including all of the Great Lakes, more than 100,000 other lakes, almost 850,000 miles of river, 92% of the Atlantic coast and 100% of the Gulf Coast.[6]

Widespread mercury advisories are not likely attributable solely (or even predominantly) to local "hot spot" sources. Instead, there is evidence that mercury emissions are chemically converted and distantly transported in the atmosphere before deposition into distant waters where bioaccumulation can occur.[3]

Several studies and reports from the USA and Great Britain have challenged the wisdom of some fish advisories. One such study concluded that the beneficial effects of fatty acids in fish outweigh any potential harmful effects of methyl mercury.[7] Methyl mercury chloride has been assumed to the principal toxic form of mercury in fish tissue. One study has complicated assessment of the health significance of mercury in fish by questioning the assumption that fish tissue mercury is predominantly in the form of methyl mercury chlorides.[4] These researchers concluded that methyl mercury cysteine is the predominant form, a chemical reported to be 20 times less toxic than methyl mercury chloride.[8]

The FDA "Action Level" for mercury in fish tissue is 1 PPM, established to limit consumers' exposure to methyl mercury to levels 10 times below the levels associated with adverse human health effects. The FDA level does not distinguish between the chloride and cysteine forms of methyl mercury, but this distinction could have a major bearing on the interpretation of fish tissue mercury levels.

Another factor in the evaluation of mercury risk is the complicated airborne transport of mercury emissions. Such transport is affected by chemical form. Gas phase elemental mercury (Hg^0) can be transported hundreds of miles before deposition. Mercury emissions from combustion sources are predominately divalent Reactive Gaseous Mercury (RGM) and particulate mercury (Hg^P). Because of the solubility and surface reactions of combustion-related mercury emissions (RGM and Hg^P), they are considered more likely to have shorter atmospheric lifetimes (quicker deposition) than elemental mercury.

For the Ohio River Valley, it has been reported that wet deposition of RGM is the primary form of returning mercury air emissions from coal-fired power plants to ground level in the Ohio River Valley, accounting for as much as 35-60% of deposition, with global background contributing the remainder.[9]

Research has shown, however, that RGM is rapidly reduced to elemental gaseous Hg^0 in the atmosphere.[10,11,12,13] After reduction, airborne Hg^0 can travel long distances before it is re-oxidized to RGM or Hg^P in the atmosphere and again becomes subject to wet (or dry) deposition. To the extent that RGM emissions convert to Hg^0 and then have to be re-oxidized to RGM to be soluble, the mercury deposition pattern extends considerably.

Several studies of mercury transport have been conducted in order to determine patterns of deposition, especially from power plants and other combustion sources.

One study, based in Steubenville, OH, concluded that deposition near the sample site was due to local and regional sources [underline added for emphasis].[14] In this study, sources were as distant as 100 kilometers away and the deposition was based solely on wet deposition of RGM.

Other studies have agreed that wet deposition is the primary form of mercury deposition but have not concluded that local or even regional sources predominate. For example, a study conducted by the Brookhaven National Laboratory covered a 10-year period.[15] They took soil/vegetation samples around three coal-fired plants, looking for evidence of mercury hotspots. They concluded that estimated increases in soil concentration suggested that less than 2% of the locally emitted mercury deposited close to these plants. Their overall conclusion was that, at least for the three power plants investigated, there was no evidence that local soil/vegetation levels of mercury were primarily due to local emissions.

Somewhat counter to the Brookhaven study, eight years of water sampling conducted by Penn State University revealed mercury concentrations higher in one Pennsylvania town than another.[16]

The site with the higher concentration of mercury was closer to upwind power plants than the other site. Interestingly, however, the downwind location with the higher mercury levels did not show any reduction in the measured amount of mercury even after significant (47%) reductions in mercury emissions from three large upwind power plants were made during the study period.[17]

One study has attributed high wet deposition not just to the immediate downwash of RGM emissions from local sources but also to the scavenging of upper altitude RGM, suggesting cross-regional transport of mercury.[18] The same study estimates that scavenging of high altitude RGM accounts for over half of the total wet deposition within the USA. Finally, the study concluded that American anthropogenic emissions contribute only approximately 20% of the total mercury deposition in the USA.

One report has identified the formation of five mercury contamination hot spots in New England and Canada (Nova Scotia), based on fish and wildlife sampling.[19] The report concludes that approximately 40% of mercury disposition is "local" to coal-fired power plants, presumably leaving 60% to non-local sources. The authors point out that factors other than air emissions (such as topography and surface water management practices) also affect biotic mercury concentrations. Interestingly, the mean concentrations of mercury among the fish tissue samples, even in the reported hot spots, were all below the FDA Action Level (1 PPM).

Despite varying estimates of mercury conversion and deposition patterns, some common themes can be identified. Mercury emitted as RGM can be "scrubbed" by precipitation and wet deposited downwind of a source. The relative contribution of such immediate deposition versus large range transport/conversion is still in doubt, but rapid localized (hot spot) deposition does not appear to predominate. Also, the health effects of mercury deposition are not immediate; so episodes of wet mercury depositions are not as important as long-term average deposition (both wet and dry), which depends on the complex chemistry of atmospheric conversions and transport. Regional controls (as via cap-and-trade) appear at least as likely as local controls to reduce fish tissue levels of mercury.

Because of the complexity of mercury's atmospheric reactions and food chain-related health effects, it is even less likely than other toxic air pollutants to create health risk hotspots.

Taking into account both the complexity of mercury pollution and the successes of the cap-and-trade approach (see Chapter 10) in other regulatory programs, (ozone and SO_2 control for example), application of the cap-and-trade approach to mercury emission reduction would make sense for reducing regional and national biotic mercury levels and would be unlikely to result in new or more intense hot spots.

Nevertheless, and despite the apparent wisdom of using a cap-and-trade approach for mercury, the U.S. EPA did not employ a cap-and-trade approach when it issued its National Emission Standard for Hazardous Air Pollutants (NESHAP) for power plants in 2011.[20]

This standard is known in "shorthand" as the Mercury and Air Toxics Standard (MATS) regulation.

Although it was aimed primarily at mercury, it also addressed other air emissions from coal and oil-fired Electric Generating Units (EGU's).

The MATS rule has been challenged in court fights up to and including the US Supreme Court.[21,22] It is a conventional "command and control" regulation requiring affected EGU's to reduce their mercury (and other) emissions to specified emission rates pro-rated based on electric power production.[23]

The basis for many of the court challenges was the cost/benefit calculation methodology used by U.S. EPA to justify regulation of mercury.

The U.S. EPA cost/benefit analyses have been described by one otherwise rule-supportive reporter as "mathematical legerdemain."[24]

The principal complaint was that, in calculating health benefits, U.S. EPA included "co-benefits" from many other air pollutant emission reductions beyond mercury.

In 2020, U.S. EPA withdrew their economic justification for the MATS rule but left its emission limit standards intact. In effect, U.S. EPA revised its economic calculations, concluding that the costs outweighed the benefits enough to reverse their previous judgement that MATS was "appropriate and necessary."

In the time between initial adoption of the MATS rule and the 2020 U.S. EPA action, actual EGU mercury emissions had already declined by 85%.[25] The rule had required compliance by 2016, so most utilities had already spent the money and installed controls by 2020.

Also, as noted elsewhere in this book, the widespread conversion from coal/oil to gas-firing greatly reduced the mercury content of the power plant fuels. The mercury reduction target from EGU's had been 91%. So, by the time that the rule's economic basis was modified by U.S. EPA, most of the targeted reduction had already occurred.

In an interesting reversal of roles, several electric utilities that had already invested (reportedly almost $20 billion) in mercury controls prior to 2020 came out against U.S. EPA's withdrawal of the MATS legal justification because it could affect their ability to recoup their costs.

If they could no longer justify electricity customer rate increases based on a regulatory mandate (the MATS rule), the approval of their rates by state utility commissions could be jeopardized.[25] In the words of the Director of Air Quality for one major utility: "The new rule could actually cost companies more money." [24]

It is interesting to ponder whether the mercury reductions could have been accomplished more efficiently if U.S. EPA had taken a cap-and-trade approach. Was it the best use of $20 Billion to insist on controlling mercury emissions beyond what could be justified by cost-benefit analysis?

Zero Release Controls

Enhanced control of VOCs and methane from natural gas operations has become the subject of regulatory proposals at both the federal and state levels.[26]

One such operation is "pigging." "Pigging" is a term for cleaning the internals of pipelines. It is employed at oil/gas extraction, transport, and processing facilities.

The process consists of introducing a rotary scrubbing device (a "pig") into the piping, then feeding the pig through the pipe to clean deposits from its internal surfaces.

Because it often occurs in outdoor, frequently remote, locations, pigging is not always a well-controlled and clean procedure.

As the pig enters and exits a pipeline, there is exposure to outside air, potentially allowing uncontrolled release of pipeline contents (oil/gas) into ambient air.

Source for methane leakage figure: Stanford University/Science,2/13/2014; Creative Commons license

If completely uncontrolled, pigging can be responsible for releasing air pollutants such as methane and VOCs.

Driven either by self-interest (escaped oil or gas is, by definition, escape of saleable commodities) or environmental regulation, many—if not most—oil/gas companies have instituted improved practices to capture (rather than release) organics from pipelines. It has been reported that modern pigging practices capture up more than 90% of these organics.[27]

For some regulatory agencies, capture of a high percentage of organics (especially methane) is not sufficient.

There have, for example, been regulations drafted to achieve 100% control of methane release from pneumatic controllers at natural gas processing plants.[28] Even ignoring the impossibility of controlling 100% of anything (enforcing a 0% allowance?) it has never been demonstrated that controlling the gap from say 90% to 100% will yield any environmental, health or even financial (by methane recovery) benefit.

Requiring 100% control of almost any type of emission is a textbook example of regulatory overreach.

Semi-Automatic Warning Labels

When purchasing almost anything—from toys to pesticides—a warning label often advises that the product contains a substance that can have adverse health effects.

For genuinely hazardous items, such warnings make sense. It is certainly appropriate to warn consumers to be careful about the use of some products.

In all too many instances, however, the warnings that are issued are not issued to inform or protect the public against a risk, but are issued to protect the seller of a product, provider of a service, or owner of a property against frivolous lawsuits.

The warnings have become so pervasive that the public can not differentiate between a warning that has some risk worth considering versus a warning on something that has virtually no risk whatsoever.

Source: Wikipedia, P. Pelletier, 8/3/2014

HOLD IT!

The best example of warnings gone awry is the consequence of California's Safe Drinking Water and Toxic Enforcement Act, better known as Proposition 65. Proposition 65 is a ballot initiative, approved by California voters in 1986, which was intended to inform and protect the public against exposure to chemicals that cause cancer or reproductive toxicity. It is with little exaggeration to say that one cannot go anywhere, or purchase much of anything, in California without being confronted with a Proposition 65 warning. A product as unlikely to be ingested as a metal storm door latch carries a label warning that "this product can expose you to chemicals including lead, which is known to cause cancer and birth defects..."

The following "warning" is posted at the entrance to most indoor parking lots in California:

> WARNING—Entering this area can expose you to chemicals known to the State of California to cause cancer and birth defects or other reproductive harm, including carbon monoxide and gasoline or diesel engine exhaust from vehicle exhaust systems.

The severity of this warning is nearly equivalent to the warnings currently found on cigarette packages in the United States. According to the Center for Disease Control, cigarettes are responsible for an estimated 450,000 deaths per year in the United States.

When making a purchase at Starbucks, you are confronted with the following notice:

> Proposition 65 Warning—Chemicals known to the State of California to cause cancer and reproductive toxicity, including acrylamide, are present in coffee, baked goods, and other foods or beverages sold here. Acrylamide is not added to our products, but results from

cooking, such as when coffee beans are
roasted, or baked goods are baked.

The numbing effect of the plethora of these notices causes people to ignore real risks. The Exide battery recycling plant, located in the Los Angeles area, was shut down in 2014, after decades of operation, after it was found that 7,500 residential properties in the vicinity of the plant had lead levels exceeding California's lead level for residential soil. The plant had Proposition 65 warnings at its entry gates. If area residents would have paid strict literal attention to these warnings, they could have concluded that the risk of living close to the Exide plant was the same as the risk of living close to Disneyland—an exaggeration of one risk and an underestimate of another.

The Los Angeles Times, in writing a piece on Proposition 65 in July 2020, wrote that "the profusion of warnings has subverted Proposition 65 and left Californians, and increasingly anyone who shops online, over warned, under informed and potentially unprotected...and it has funneled hundreds of millions of dollars into a handful of clients and their repeat clients." [29] In this sentence, "their" means the client's lawyers.

Taken to an extreme, one could file a lawsuit against the State of California for failure to post Proposition 65 warnings at all points of entry into California and at conspicuous locations throughout the state. Proposition 65 requires that:

> *...no person in the course of doing business shall*
> *knowingly and intentionally expose...a chemical*
> *known to the state to cause cancer or reproductive*
> *toxicity...without first giving clear and reasonable*
> *warning."*

The State of California's far-reaching operations are clearly exposing California's 41 million residents and visitors to such air and water borne chemicals emitted by the State of California's road maintenance and other operations. Proposition 65 not only empowers the state attorney general and local district attorneys to file suits, but also "any

HOLD IT!

person in the public interest."

---

[1] EPA, *US Army Repeals 2015 Rule Defining Waters of the United States*, US EPA, September 2019

[2] Forney, S.Z., *Final Navigable Waters Protection Rule Established*, A&WMA Zephyr, June 2020

[3] Duckett, E.J., *Targeting Air Emissions: Are There Hot Spots*, Proceedings of the Air & Waste Management Association Annual Conference and Exhibition, June 2011

[4] *2007 Ambient Air Quality Monitoring and Emission Trends Report*, Pennsylvania Department of Environmental Resources, Harrisburg, PA

[5] *Mercury Compounds-Hazard Summary*, USEPA, Technology Transfer Network, Washington, DC, January 2000

[6] US EPA, *National Listing of Fish & Wildlife Advisories*, Technical Fact Sheet, 2007

[7] Hibbeln, J., Davis, J., Steer, C. et al.; *Maternal Seafood Consumption in Pregnancy and Neuro Developmental Outcomes in Childhood: An Observational Cohort Study*; Lancet 2007; pp. 578-585

[8] Harris, H., Pickering, I. and George, G., *The Chemical Form of Mercury in Fish*, Science, August 29, 2003

[9] Crist, K., Kim, M and Lin, P., *Evaluation of Transport and Deposition of Mercury from Coal-Fired Power Plants Using 3-D Chemical Transport Model*, Proceedings of Air Quality VII An International Conference on Carbon Management, Mercury, Trace Substances, SO_2, NOx and Particulate Matter, Arlington, VA, October 26 – 29, 2009

[10] Danilchik, PA., Imhoff, R., et al., *A Comparison of the Fate of Mercury in Flues and Plumes of Coal-Fired Boilers*; Presented at the Mercury as a Global Pollutant Conference; Whistler, British Columbia, July 10-14, 1994

[11] Laudal, D., and Prestbo, E., *Investigation of the Fate of Mercury in a Coal Combustion Plume Using a Static Plume Dilution Chamber*; U.S. Department of Energy; Energy & Environmental Research Center; Grand Forks, ND, October 2001

[12] Edgerton, E., Hartsell, B. and Jansen, B., *Field Observations of*

85

Mercury Partitioning in Power Plant Plumes; Proceedings of the International Conference on Air Quality III: Mercury, Trance Elements and Particulate Matter; Arlington, VA, September 2002

[13] Landis, M., Ryan, J., et al., *Plant Crist Mercury Plume Study*; Proceedings of Air Quality VII: International Conference on Carbon Management, Mercury, Trace Substances, SOx, NOx and Particulate Matter; Arlington VA, October 26-29, 2009

[14] Keeler, G., Landis, M. et al, *Sources of Mercury Wet Deposition in Eastern Ohio, USA*, Environmental Science and Technology, November 19, 2006

[15] Sullivan, T., Bowerman, B. et al., *Local Impacts of Mercury Emissions from Coal Fired Power Plant*; Brookhaven National Laboratory; Upton, NY, March 2005

[16] *Toxic Mercury Pollution*; PA Dept of Environmental Protection, 2006

[17] Biden, D., *DEP Data Shows Power Plant Mercury Emissions Do Not Cause Local Hot Spots*; PA Electric Power Generation Association, June 5, 2006

[18] Selin, N. and Jacob, D.; *Seasonal and Spatial Patterns of Mercury Wet Deposition in the US: Constraints on the Contribution from North American Anthropogenic Sources*; Atmospheric Environment; Vol. 42 pp 5193-5204, 2008

[19] Evers, D., Han, J-J, *Biological Mercury Hotspots in the Northeastern US and Southwestern Canada*; Bioscience; Volume 57, No.1, January 2007

[20] US EPA, *Proposed Rule: National Emission Standards for Hazardous Air Pollutants from Coal-and-Oil Fired Electric Utility Steam Generating Units and Standards of Performance for Fossil Fuel Fired Electric Utility, Industrial-Commercial-Institutional and Small Industrial-Commercial-Institutional Steam Generating Units*, 40 CFR Parts 60 and 63, March 16, 2011

[21] Environmental and Energy Law Program (EELP), *Mercury and Air Toxic Standards*, Harvard University Law School, September 28, 2020

[22] *History of the Mercury and Air Toxics Standards (MATS) for Power Plants Regulation*, US Environmental Protection Agency

(EPA), September 23, 2020

[23] *Mercury and Air Toxics Standards for Power Plants*, US EPA, July 17, 2020

[24] Friedman, L and Davenport, C, *EPA Weakens Controls on Mercury*, The New York Times, December 7, 2020

[25] Proctor, D., *EPA Nixes Legal Justification for MATS Rule*, Power, April 16, 2020

[26] *Oil & Natural Gas Sector Emissions in New York*, New York State Department of Environmental Conservation, November 8, 2018

[27] *Vapor Recovery and Gathering Pipeline Pigging*, Lessons Learned from Natural Gas Star, US EPA, July 23, 2008

[28] *Draft Proposed Rulemaking: Chapter 121 – General Provisions Article III*. Air Resources, PA Department of Environmental Protection, April 11, 2019

[29] *You See the Warnings Everywhere. But does Prop. 65 Really Protect You?*, Los Angeles Times, July 23, 2020

E. Joseph Duckett and Jeffrey L. Pierce

CHAPTER 7: MARKETING PROPAGANDA/GIMMICKRY

HUGGING A TREE has become good marketing. Products as diverse as cigarettes, potato chips and straws can get a marketing boost by aligning with a perceived environmental benefit.[1] This tends to dilute the credibility of environmental protections and co-opt legitimate objectives for commercial gains.

Here are some examples...

Green Cigarettes

Under the heading "Natural Tastes Better," Natural American Spirit cigarettes are advertised as containing premium natural tobacco grown in a "responsible, <u>sustainable</u> way through our <u>earth-friendly</u> and <u>organic</u> growing program." The ad continues with: "we strive to reduce our <u>footprint</u> on the earth by using <u>recycled</u> materials and renewable energy sources like <u>wind</u> power. Protecting the earth is as important to us as it is to you."[2] [Underlines added for emphasis.]

Quite a few environmental buzzwords packed into a single ad. Towards the bottom of the ad, there is a note that: "No additives in our tobacco does not mean a safer cigarette." Promoting cigarette smoking by appeals to environmental stewardship seems highly hypocritical.

Source: Flickr, 3/2/2008; Creative Commons license

Paper Straws/Plastic Pollution

In response to reports of extensive oceanic plastics pollution, alarms have been raised about the use and disposal of plastic straws. In the USA and elsewhere, restaurants have proudly announced that they are banning the use of plastic straws. In their place, they now offer only straws made of paper—or, in some cases, no straws at all.

Leaving aside the pros and cons of plastics versus paper (oil versus trees as raw materials; comparative energy consumption of plastic versus paper), what exactly is the likely environmental impact of substituting paper for plastic straws?

Source: kapgar.com , 8/10/2019; Creative Commons license

Most of the reports of extensive oceanic plastic pollution have focused on conditions along the ocean shores of developing countries. For example, a recent report, widely referenced as the stimulus for plastic straw banning documented plastic pollution directly off the coast of the Philippines.[3]

Among the top 10 plastic waste dischargers to oceans, 8 are Asian countries; 2 are in Africa. The USA is ranked as No. 20 on this list, accounting for less than 1% of total plastic waste discharge to oceans.[4]

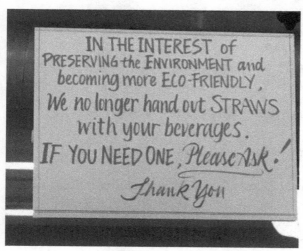

Source: E.J. Duckett photograph 8/27/2020

Since straws are only about 1% of all USA-based plastic wastes, the elimination of all plastic straws in the USA would only reduce the total global ocean discharge of plastics by less than 0.01%.

For the USA and Europe, the vast majority of all

wastes are collected/disposed in controlled processes—landfilling or incineration—rather than being mismanaged and dumped into the ocean.[5] Banning plastic straws will obviously reduce the amount of plastic straw waste in landfills but, as noted above, straws represent less than 1% of all plastic disposal in the USA.[6] Further, on the related subject of microfibers as oceanic pollutants, a recent survey of 617 locations worldwide has concluded that less than 10% of the collected fibers were synthetic plastics; 92% were natural, mainly cellulosics.[7]

If straws are such a small component of municipal wastes and the vast majority of the wastes are disposed on land, the oceanic impact of a USA plastic straw ban seems minimal. Claiming environmental achievement by banning straws is eco-gimmickry with little or no real impact. If the bans were imposed in developing countries, there could be a benefit, but these countries are usually the least able to afford more expensive straws—or the added cost of proper disposal.

Marketing Adjectives

To profess the environmental superiority of their products, some manufacturers have begun advertising them with such adjectives as sustainable, eco-friendly, biodegradable, organic, recyclable, free-range and green. Unfortunately, few of these terms are well-defined and even fewer are subject to governmental standards.[8]

Imagine trying to rate or define the sustainability of a product. Would it be based on raw material use, energy efficiency, emissions release, recycled content or some complex combination of them all? How would they be weighed by relative importance? What baseline would be used for judging a sustainable versus "normal" product? For multi-component products, like

Source: Open Clipart, SVG ID 54307, 12/26/2016; Creative Commons license

91

computers, would there be a blended or weighted sustainability or recyclability rating?

In the absence of recognized standards, touting the eco-friendliness of a product is essentially risk-free and too often truth-free as well.

Zero Discharge or Zero Emissions

Among the phrases that elicit positive eco-vibes, "zero discharge" or "zero emissions" are near the top of the list. Unfortunately, the underlying evidence to which these phrases are applied has to be closely examined.

In keeping with the laws of mass and energy conservation, matter is neither created nor destroyed. So, if a fuel is combusted or a product is manufactured, the input materials all go somewhere—either to the product or to a waste stream.

In turn, if the waste stream is further processed (e.g., wastewater treatment or air emission control), the quantity and composition of the wastes can be detoxified and minimized but the fact remains that everything has to go somewhere.

In reality, zero discharge must always be modified with adjectives like "liquid" to indicate what discharge has been eliminated.

For example, "zero liquid discharge" could apply to a waste stream which, after treatment, results in a solid (sludge) residue and a cleaned effluent that is evaporated.

In effect, the potential liquid discharge (effluent) would be converted from liquid to water vapor by evaporation. In this case, the effluent would be discharged as a gas rather than a liquid. One discharge is eliminated but another is formed. Not exactly zero discharge if you take all releases into account.

One of the more interesting marketing twists on zero emissions is the labeling of some vehicles as PZEV—Partial Zero Emission Vehicle. It simply means that the car has advanced emission controls that minimize exhaust components—but that doesn't mean that emissions are eliminated completely. It just sounds a lot better to be a partial zero emission vehicle rather than an almost completely controlled vehicle (ACCV), or being labeled a partially polluting vehicle (PPV). The word "zero" connotes perfect environmental performance which is positive marketing indeed. Unmentioned is that, at least in California, a PZEV is simply compliant with governing California vehicle emission standards. The vehicle manufacturer could not sell the vehicle in California unless it was a PZEV.

Source: Green Car Reports, 7/27/09; Creative Commons license

An even more egregious use of the term zero emission is Tesla's assertion that they manufacture zero emission vehicles.

It is correct to state that there are no tail pipe emissions from a Tesla.

Zero tail pipe emissions is certainly a Tesla benefit resulting from displacing internal combustion engines with electric cars in areas of non-attainment for criteria air pollutants, such as Los Angeles. However, this is far from the whole story.

Air emissions are produced in the generation of electricity; thus, a Tesla is not zero emissions.

The actual value of Tesla air emissions reduction depends on: 1) on the fraction of renewable energy which is generated on the electric power network from which the Tesla car is drawing its electricity; 2) the type of fossil fuel that is used to generate the non-renewable power (coal versus natural gas) on that electric power network; and 3) the proximity of the fossil fuel power plants to where the Tesla is driven (i.e., are the power plants in the same air basin as where the Tesla is driven?).

With the intense scrutiny on carbon dioxide (CO_2) emissions, as related to global warming, let's focus on CO_2 emissions. According to the California Air Resources Board (CARB), the carbon intensity (CI) of gasoline, when used as a vehicle fuel, is about 100 grams of CO_2 equivalent per megajoule (100 g CO_2/MJ).[9]

In 2020, about 52% of the electric power consumed in California was from fossil fuel. The balance was produced from non-fossil sources—solar, wind, nuclear and hydroelectric. The fossil fuel was 99.8% natural gas.

According to CARB, the average CI of electricity in California, as of 2020, was about 82 g CO_2/MJ. When factoring in the much higher efficiency of an electric motor versus an internal combustion engine in converting energy into mechanical power, the CI of California electricity when used as a vehicle fuel is 24 g CO_2/MJ. A Tesla yields a significant reduction in CO_2 emissions, but is hardly zero emissions.

California is blessed with electric power with a very low CI. The average CI for Missouri's electric power is about 350 g CO_2/MJ. Missouri's electric power is about 82% from coal, 9% from natural gas and 9% from renewable sources. The CI of Missouri's electricity, when used as a vehicle fuel in Missouri, is then roughly 100 g CO_2/MJ—the same CI as gasoline. Each state produces a different CI for a Tesla.

It could be argued that renewable power should be viewed as having a dedicated pathway to electric vehicles.

In other words, electric vehicles run only on renewable power alone. The CI of electricity used in the electrical vehicle should not be based on the state average electric power CI, but that on that of renewable electricity. We will accept that argument since CARB has sanctioned such theoretical pathways under its low carbon fuel standard (LCFS) program. As a side note, however, renewable power is not carbon free.

Even wind and solar energy power sources have a CI, when considering the life-cycle CO_2 impact of renewable power equipment manufacturing and renewable power delivery.

Now, having opened the door of direct energy pathways, the substitute of renewable natural gas (RNG) for compressed natural gas (CNG) as a vehicle fuel should be discussed. RNG is a natural gas equivalent that is produced from decaying organic material, including cow manure. Dozens of dairy digester gas RNG projects are now coming online in California and across the USA.

CARB typically assigns dairy digester gas RNG, when used as CNG, a CI of about -260 g CO_2/MJ. The value is negative because CARB's modelling fully considers the capture of methane at the farm that supports the RNG production facility.

Without the RNG production project, that methane would be directly emitted to the atmosphere. The RNG produced by the RNG production facility is injected into the natural gas pipe network and it finds its way to CNG fueled vehicles.

The tracking concept is similar to the theoretical pathway that was described above to justify calling an electric car zero emissions (or low emissions).

In the electrical vehicle case, the renewable electrons entering the electric power distribution system are tagged and find their way to an electric vehicle. In the RNG vehicle case, the RNG is tagged from the point of production through existing natural gas pipelines to the point of use in CNG vehicles.

But unlike the electric vehicle using a fuel with a CI of 24 g

CO_2/MJ, the fuel used in the CNG vehicle has a CI of -260 g CO_2/MJ.

Another valuable form of RNG is RNG produced from landfill gas. CARB typically certifies the CI of RNG produced from landfill gas for use as CNG vehicle fuel to be in the vicinity of 40 g CO_2/MJ. An electric vehicle in California betters this CI, but not by much, and both RNG fuel's CIs are substantially better than gasoline at about 100 g CO_2/MJ.

As of 2021, 90% of the CNG used in California as vehicle fuel was RNG. CARB calculated that the average CI of that CNG was -0.9 g CO_2/MJ. In other words, the buses, garbage trucks, tractor trailers and fleet vehicles in California are, as whole, running on a slightly negative CI fuel.

The CI of California's CNG portfolio will drop, as more dairy digester RNG projects come online. New dairy digester RNG will displace existing landfill gas RNG now in the CNG portfolio. The displaced landfill gas RNG will be sold as CNG to markets in other states—states with higher electric power CIs than California.

As explained above, the California CNG market is fully committed to RNG, and can absorb no additional RNG. It would seem that California would be aggressively promoting the expansion of the use of CNG as a vehicle fuel—by incentive or by regulation.

Most existing large trucks (e.g., tractor trailers) can easily, cost effectively and relatively quickly, be converted to CNG. The conversions would create demand pull for RNG.

Instead, California is going "all in" for electric vehicles. It will require decades to replace the California large truck fleet with electric trucks. The bulk of the existing large trucks can be converted to CNG in years, not the decades needed to convert to an all-electric fleet.

The purpose of the above discussion is not to argue that RNG CNG fueled vehicles are better than electric vehicles. The purpose is to show that the carbon math has clearly not been thought through, at least in the short term, by those aggressively promoting and subsidizing electric vehicles.

In the ultimate irony, several public transit fleets are converting some of their existing CNG fueled buses to electric buses. The public transit decisionmakers seem to be caught up in the "electrification euphoria" sweeping California, but they are actually switching to a higher CI fuel.

Further complicating the adoption of electric vehicles as zero emission alternatives to conventional cars is the unavoidable link between electric automobile production and raw material extraction. As noted in a 2021 New York Times article, "electric cars may not be as green as they appear." [10]

Electric cars require batteries which, in turn, require lithium. There is only one large scale lithium mine in the USA and it produces only 2% of the world's supply.

The USA has been reliant on imported sources of lithium, a vulnerability that could limit the expanded deployment of electric vehicles, and also the growth of solar power and wind power—both of which will require large scale power storage.

Plans for opening new lithium mines in the USA are in the works. As noted by the Times, however, mining of lithium and other raw materials essential to all-electric technologies "are often ruinous to land, water, wildlife and people."

Thus, we come back to the theme of this book—Hold It!

Today's prevailing assumption, at least in California, is that anything electric is good, and that anything that involves combustion is bad.

The assumption seems to ignore that fossil fuel must play a role in power generation for a very long time and ignores that there are other low CI combustion fuels available.

The absolute embrace of "electric is good and combustion is bad," without thinking through all aspects of this simplification—actual quantification of environmental performance, possible alternative low carbon paths or co-paths, financial and social impacts of any plan, implementation time required, and short-term actions versus long term actions—is a rush to judgment, and an oversimplified response to a complex problem.

E. Joseph Duckett and Jeffrey L. Pierce

Automatic Bill Paying to Save the Earth

Anyone who pays periodic bills the old-fashioned way, by mailing a personal check, has been urged to take the environmentally friendly step of paying by automatic withdrawal from their bank account. This is promoted as saving paper (envelopes, checks), energy (mail delivery) and cost (postage). In reality, automatic paying has obvious advantages to the merchants (especially credit card companies) who no longer need to send bills or worry about getting paid on time.

Source: Wikimedia Commons,5/23/2017; Creative Commons license

Source: PNGing.com,12/28/2018; Creative Commons license

In this case, the environmental benefits are negligible, but payers are made to feel eco-friendly about paying automatically.

Eco-guilt can make life easier for merchants seeking payment.

BPA

Biphenyl A (BPA) is a synthetic organic compound used in the manufacture of polycarbonate and other plastics. Polycarbonate plastics are commonly used in food and beverage containers, including as liners for metal containers.

BPA-containing plastics are typically clear, tough and shatter resistant—all important properties for containers.

Because of its presence in food containers, BPA has attracted concerns for its potential health effects.

HOLD IT!

Source: M. Morgan in flickr,7/21/2013; Creative Commons license

A 2003 analysis from the U.S. Center for Disease Control (CDC) reported the detectable presence of BPA in 93% of more than 2,000 randomly selected urine samples from the USA.

As with many such findings of "detectable" compounds, the question becomes whether such detection represents a legitimate health threat—or not.

Alarming warnings have been issued by some researchers and environmental/health groups.[11] They note, correctly, that BPA is known to leach from plastic containers, especially when heated. They have also noted, again correctly, that BPA behaves like estrogen and can bind to the same receptors as natural female hormones.

In 2012, the U.S. Food & Drug Administration (FDA) banned the use of BPA in baby bottles and infant formula packaging. Some individual states have banned BPA in other food and beverage containers.

The BPA restrictions are clearly cautionary but there are indications that they may be examples of regulatory overreach. Testing of the "fate" of BPA in humans has shown that BPA (and its byproducts) are readily excreted and are not retained in the body for long. Further, a 2018 FDA lab study concluded that, among BPA-dosed laboratory rats, "minimal" effects were found.[12]

The FDA added that "currently authorized uses of BPA are safe for consumers."

Obviously, consumers are free to avoid BPA-containing plastics if they so choose, but regulatory bans are exaggerated reactions to what appear to be minimal risks. Advertising bottles as "BPA-Free" may have marketing punch but very little real health protection impact.

---

[1] Fehrenbacher, J., *Top 5 Dumbest Greenwashed Earth Day Gimmicks, Inhabitat,* April 2013

[2] Advertisement for Natural American Spirit Cigarettes

[3] Martin, M.G., *Philippines Third Biggest Source of Plastic Pollution in World's Oceans,* Philippines Lifestyle News, September 24, 2017

[4] Jambeck, J. et al., *Plastic Waste Inputs from Land into the Ocean,* Science, February 13, 2015

[5] Borenstein, S., *Science Says: Amount of Straws, Plastic Pollution is Huge,* AP News, April 20, 2018

[6] Bufkin, E., *Why Plastic Straws are Actually Better for the Planet than Paper Straws,* The Federalist, September 17, 2019

[7] Suaria, G., et al, *Microfibers in Oceans Surface Waters: A Global Characterization,* Science Advances, June 2020

[8] Graber-Stiehl, I., *How Bogus Eco-Friendly Products Trick You,* Gizmodo, August 2018

[9] California Air Resources Board (CARB)—https://ww2arb.gov

[10] Penn, I and Lipton, E., *The Lithium Gold Rush: Inside the Race to Power Electric Vehicles,* The New York Times, May 6, 2021

[11] Schroeder, M., *Is it Safe to Drink from Plastic Bottles,* US News & World Report, 2017

[12] Pitts, N.M., *The BPA Controversy,* Fisher Scientific Science News, 2018

CHAPTER 8: MYTHS

Cancer is Worse Now

FROM READING SOME alarmist literature, the impression is that cancer rates in the USA have grown over the past few decades.

In fact, the opposite is true. In the Center for Disease Control's Annual Report to the Nation on the Status of Cancer,[1] it is confirmed that "cancer rates continued to decline (2001-2017) in the USA for all cancer sites combined." Further, the decreases crossed all major racial, ethnic and age groups.

The overall cancer death rates decreased by an average of 1.5% per year. Lung cancer death rate, in particular, decreased almost 5% among men over the period 2013-2017, although lung cancer continues to be the leading cause (about 25%) of all cancer deaths.

Interestingly, the rates of cancer incidence (not necessarily deaths) decreased among Black, Native American and Hispanic men.

Figure 14
All Cancer Sites Combined

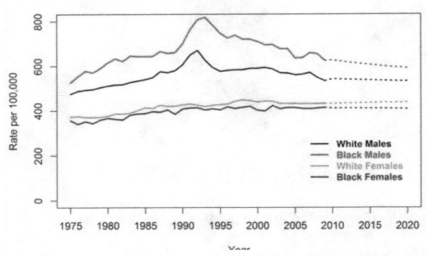

Source: National Library of Medicine, National Institutes of Health, Cancer,
2/3/2015

All of this progress is despite the acknowledgement that national goals for reduction of cigarette smoking, alcohol abuse and obesity—non-environmental factors—have not been met.

Pollution is Worse Now

Similar to the literature on environmentally related risks, it has been popular to contend that "pollution" is getting worse. Again, as with cancer rates, the opposite is true.

The University of California-Berkley has conducted a very thorough analysis of water pollution in the USA over the years since the 1972 passage of the Clean Water Act.[2]

The team analyzed data from 50 million water quality measurements collected at 240,000 monitoring sites. Their conclusion was that most of the 25 wastewater pollution indicators showed improvement—not degradation. Example indicators were dissolved oxygen concentrations, fecal coliform bacteria contamination and fishability.

Although, at least for water pollution, the researchers concluded that the costs of the cleanup have exceeded the benefits, their bottom-line conclusion of improved water quality strengthens the record versus water pollution alarmists.

Turning to air quality, the improvements have been widespread and continuous.[3] Unlike the water pollution case, the benefits of air pollution control have been shown to outweigh the costs.

Figure 15
Changes in Air Pollution Since 1970

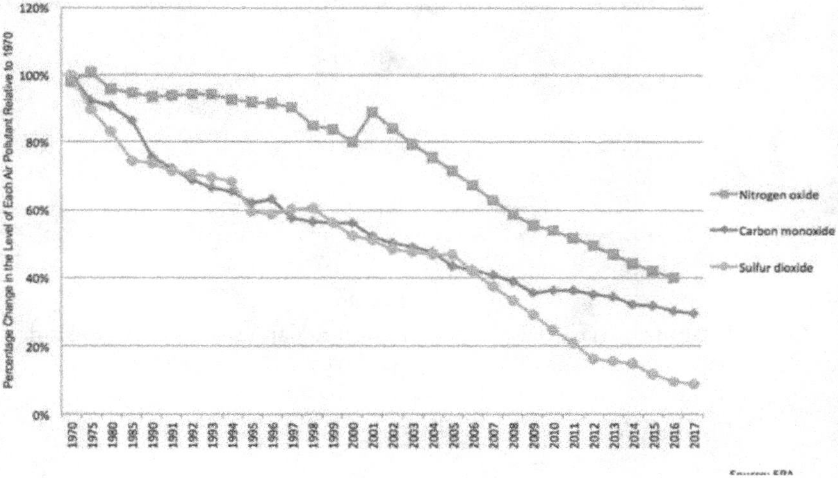

Figure 15 above from the U.S. EPA depicts the general downward trend (improvement) in air pollutants across the USA.

Despite such national progress, over the past few years, there has been a wave of articles and publications casting negative views on air quality in many specific regions of the country. One such area has been Allegheny County, PA, which includes and surrounds the City of Pittsburgh. Among these negative views have been a series of annual "State of the Air" reports of the American Lung Association (ALA). Over the years, the ALA Report has reinforced the image of Pittsburgh

E. Joseph Duckett and Jeffrey L. Pierce

as the "Smoky" City by presenting a very exaggerated and unbalanced picture of the actual air quality within the Pittsburgh area. Figures 16 and 17 contradict this view, at least for particulates.

Figure 16
Pittsburgh Area PM$_{2.5}$ Annual Weighted Means by Year
2000 to 2017

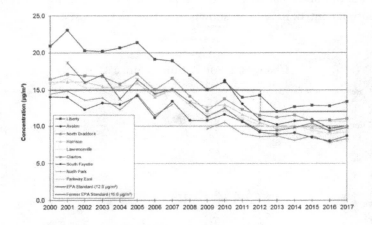

Figure 17
Pittsburgh Area PM$_{2.5}$ Annual Design Values by 3-Year Period
2000 to 2017

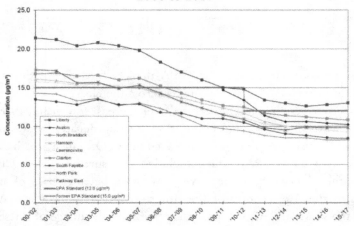

HOLD IT!

The most recent ALA Report assigns Allegheny County an "F" grade and asserts the County "remains a long way from being healthful." [4]

In reaching its conclusions, the ALA cherry-picks data from individual monitoring sites and then assumes that these sites represent not only Allegheny County but the entire PA-OH-WV region from New Castle, PA to Weirton WVA.[5] The Lung Association's methodology has been exposed in the past as completely unscientific and misleading.[6]

Probably because of the Pittsburgh area's long history of air pollution concerns, Allegheny County is one of the most heavily monitored regions of the country.

As a result, testing data from multiple sites can be reviewed to identify variations in air pollutant concentrations from one area to another within the county.

At present, the local regulatory agency maintains a total of 17 air pollution control monitoring stations, each with a different mix of measurement equipment.

These are listed in the table below. In addition to these monitoring stations, specialized measurement studies have been conducted for some areas.

At least from an air pollution perspective, there is no single "Allegheny County;" instead there are several air pollution regions with the County.

By almost all accounts (including those who have authored the negative reports), progress on improving air quality in Allegheny County has been dramatic and positive.

There are no pollutants for which the County is in worse shape today than it was twenty years ago.

Having said this, it is worthwhile reviewing the situation on a pollutant-by-pollutant basis.[7]

E. Joseph Duckett and Jeffrey L. Pierce

Table 2—Allegheny County Health Department
Air Quality Monitoring Locations

Locations	Measured Pollutants					
	Ozone	PM	SO$_2$	CO	NOx	Toxics*
Avalon		■				■
Clairton		■				
Collier		■				
Flag Plaza				■		
Glassport		■				
Harrison	■				■	
Lawrenceville	■		■			■
Liberty		■	■			
Lincoln		■				
Manchester		■				
Natrona		■				
North Braddock		■				
North Park	■	■				
Parkway East				■	■	■
Russellton		■				
South Fayette	■	■				
West Deer		■				

*Includes H$_2$S, Pb and designated Hazardous Air Pollutants
Source: Allegheny Count PA Health Department, 2019

Particulate Matter (PM). In its "State of the Air" report, the ALA lumped Pittsburgh with the entire PA-OH-WV Region from Weirton, WV to New Castle, PA, and claimed that this area was the eighth most polluted area in the country for short-term particle pollution (so-called PM$_{2.5}$ or "fine" particulates). "Fine" particulates are associated with significant health effects because they are small enough to be drawn deeply into the inner reaches of our lungs. The ALA rating was based on individual monitoring sites not at all representative of the entire region.

The USA annual ambient standard for PM$_{2.5}$ is 12 micrograms per cubic meter ($\mu g/m^3$) on a 3-year averaged basis. Over a recent 3-year period, only one of the eight monitoring stations in Allegheny County measured an average concentration above this standard. Even the one out of compliance Station averaged only 12.2 $\mu g/m^3$. The trend was definitely towards <u>lower</u> concentrations.

In addition to the annual standard, there is also a federal 24-hour ambient air quality standard for $PM_{2.5}$. Until 2006, the standard was 65 $\mu g/m^3$. Since 2007 the standard has been reduced to 35 $\mu g/m^3$. For 2019, the standard was again exceeded at only one of the eight monitoring stations in Allegheny County. That station only exceeded the 24-hour standard on a total of nine days. 97% of the concentrations at this station were at or below the standard. By any evaluation, the trend has again been downward, with the vast majority of the stations (7 out of 8) consistently at or below the U.S. EPA standard.

<u>Sulfur Dioxide (SO_2)</u>. Sulfur Dioxide is a long-recognized air pollutant primarily associated with industrial areas, including power plants. The former federal ambient air quality standards were 0.03 Parts Per Million (PPM) annually and 0.14 PPM as a 24-hour average. Beginning in 2010, a new 1-hour federal standard of 75 Parts Per Billion (PPB) (0.075 PPM) took effect, essentially replacing the annual and 24-hour standards.

Among the five monitoring stations for SO_2 in Allegheny County, <u>none</u> exceeded either the former annual or 24-hour standards in 2019. Two stations recorded exceedances of the 1-hour standard. There were five hourly exceedances at one of these stations, therefore, only 0.06% of the total annual hours at one site were in excess of the 1-hour standard. There were only two hourly exceedances at the other exceedance site. As with most of the other pollutants, the historical trend for SO_2 is downward at all stations.

<u>Carbon Monoxide (CO)</u>. Carbon Monoxide is a well-recognized toxic air pollutant for which federal hourly and 8-hour standards have been in place for several decades. In recent years, there have been <u>no</u> exceedances of either the 1-hour or the 8-hour CO standards at any of the three CO monitoring stations in Allegheny County. In fact, the County has not exceeded the 8-hour standard since 1987.

<u>Nitrogen Oxides (NOx)</u>. Nitrogen Oxides are considered a health threat in their own right and are also precursors for atmospheric

ozone. U.S. EPA has established both annual average and 1-hour federal ambient air quality standards since 2010. Measurements taken at the two NO_x monitoring stations in Allegheny County reveal attainment of both of these federal standards. In fact, even the 1-hour standard has been met continuously.

Lead (Pb). Lead is another well-recognized toxic air pollutant monitored within the County. Since 1991, all of the monitoring stations for lead have measured concentrations well below the pre-2009 ambient standard, which was based on quarterly averaged measured emissions. Since 2009, the standard has been reduced by 90%; it is now at 0.15 $\mu g/m^3$ on a three-month rolling average basis. Even after this significant reduction in the ambient standard, all monitoring stations within the county have complied, with measured levels below the revised standard; with exception of one site, which was subsequently corrected (reduced) in the next year.

Ozone (O_3). Ozone is considered a secondary pollutant which results from atmospheric reactions involving oxides of nitrogen (NO_x) and volatile organic compounds (VOCs). The prevailing federal ambient air quality standard is 0.070 PPM based on maximum 8-hour averages within a calendar day. For 2019, there were no exceedance days for the 8-hour ozone standard. Since 1997, average 8-hour ozone concentrations have been consistently reduced throughout the county.

Toxic Air Pollutants. There are no actual federal ambient air quality standards for toxic air pollutants. There have been, however, several toxic air pollutant studies conducted in Allegheny County over the past decade. These include studies conducted by the Allegheny County Health Department, Carnegie Mellon University, and the University of Pittsburgh. Because there are no official ambient standards, it is difficult to put perspective on the measured concentrations. One approach is to calculate "acceptable" ambient levels based on reported toxicity studies and standardized air dispersion modeling. Another approach is to compare measured concentrations with levels measured in other areas of the country. Based on these studies, several broad conclusions can be drawn, namely:

- For most areas of Allegheny County, the levels of measured toxic air pollutants are unremarkable in the sense that they fall below a cancer risk level of 1 in 100,000 over a lifetime of exposure (1×10^{-5}) risk.

- Only one area of the county had measured air concentrations in excess of the 1×10^{-5} threshold.

- The most significant air toxics measured to date are diesel particulates, coke oven emissions, benzene, formaldehyde, trichloroethylene and dichlorobenzene. The single highest risk was attributable to diesel emissions from mobile sources.

- As with all the other pollutants for which ambient air quality standards have been established, all concentrations of toxic air pollutants have been reduced over time. Even for the area of the county for which the highest levels of toxic air pollutants have been monitored, the calculated health risk levels, when applied to the actual population of this area (approximately 50,000 people), indicates an additional risk of less than one cancer case over a 70-year lifetime. Although it is politically incorrect to dismiss any additional cancer risk, it is also intellectually dishonest to suggest that such risk represents an alarming health threat. The background national lifetime risk of getting lung cancer (combined smokers and non-smokers) is at least 6%, which is equivalent to more than 3,000 lifetime cancer cases in a population of 50,000 people. The risk for smokers is therefore much higher.

In conclusion, air pollution conditions in Allegheny County and many locations elsewhere in the USA are nowhere near as dire as suggested by some studies and news reports. No one is suggesting that air pollution regulations should be weakened or that the efforts by government and industry to reduce air emissions are no longer necessary. Surely, however, it is time to stop bashing areas of the USA with the same tired air pollution stereotypes which have persisted for many decades.

E. Joseph Duckett and Jeffrey L. Pierce

Everyone acknowledges that many areas, including Allegheny County, had very serious air pollution problems in past years. Nearly everyone also now acknowledges that progress to clear the air has been dramatic and successful. It is time to acknowledge that success rather than to exaggerate current air quality conditions.

Recycling Relieves Material Shortages

Aside from reducing the need for landfill disposal of wastes, one of the major promotional advantages of municipal waste recycling is relieving shortages of raw materials for production of new products.

Recycling newspaper and cardboard is intended to reduce demand for virgin wood pulp thereby reducing the need for deforestation. Recycling steel cans is intended to reduce the need to mine or import iron ore and coking coal. Recycling aluminum cans is aimed at reducing the importation and refining of bauxite. Recycling glass cullet is intended to reduce the extraction of sand and soda ash. Recycling plastics is to reduce demand for oil.

Figure 18

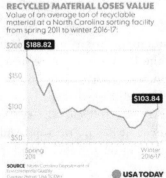

Source: pixabay,6/4/2014; Creative Commons license

110

To some extent, it is obvious that substituting waste-derived materials in a production process reduces the need for virgin raw materials. But to suggest that, for the USA, recycling has significantly relieved raw material shortages requires evidence that: a) there are material shortages; and b) that the recycled materials are used extensively in this country.

There is little reason to believe that the USA has shortages of iron ore, coke, sand or soda ash. Much of the alumina used in manufacturing aluminum is imported, so waste aluminum recycling does reduce the need for such importation. The value of recycling aluminum is reflected in the strong prices paid in the USA.

If recycled materials were needed to relieve homegrown raw material shortages, only minimal export would be expected. But the export of large proportions of wastepaper, cardboard and plastics—much of it to China—suggests that there is no shortage of wood pulp or any other form of pulp substitutes in this country.[8] In fact, due to China's recent reluctance to import recycled wastes from the USA, it appears that, if anything, there is actually a glut of recycled materials both in the USA and worldwide—at least for paper and plastics.[9]

Renewable Electric Power is Problem Free

Renewable energy was initially promoted for two reasons.

First, it was believed that we were headed for severe fossil fuel shortages—at least shortages of the relatively clean and portable fossil fuels—natural gas and oil.

Second, renewable energy was and is believed to be, in the balance, a relatively environmentally benign energy source.

The Harvard Business School[10] concluded in 1979 that declining reserves of natural gas would almost certainly require stern mandates to convert then existing natural gas-fired industrial boilers and electric power plants to coal, by 2005, to husband what limited natural gas would then remain. As we all now know, the natural gas supply situation in the USA has been reversed. There is currently a relative abundance of natural gas and oil in the USA. The experts had spoken, but they were wrong.

Source: pikist.com,6/6/2017; Creative Commons license

As a tongue in cheek remark, the only sources of energy that we can think of that ever came close to being exhausted were renewables.

In 16th Century England, after clearing almost all of the trees, renewable biomass was replaced by a non-renewable (coal), and renewable whale oil, after almost driving whales to near extinction, was replaced by a non-renewable (kerosene).

Because renewable power emits virtually no CO_2, renewable electric power is currently viewed as a primary tool in combatting global warming.

The waning concern of exhausting our fossil fuels has now largely given way to promotion of the environmental benefits of renewable electric power.

The benefit primarily being CO_2 reduction, since emission controls are largely in place for other air pollutants.

Renewable electric power technologies include: solar; wind; small-scale hydroelectric; large-scale hydroelectric; geothermal; tidal; wave and others. Solar and wind have been responsible for almost all of the growth in renewable electric power over the last decade.

HOLD IT!

Both solar and wind have been heavily incentivized by tax credits and by public utility renewable portfolio carve outs. Solar and wind are welcomed as largely problem-free.

There are, however, environment issues with both solar and wind power.

Wind power kills birds, it is noisy, and it is an eyesore.

In California and in other western states, solar panels blanket hundreds of thousands of acres of desert land, dramatically impacting the flora and fauna and the landscape.

Recently, the California solar industry fought the designation of the Joshua Tree as a threatened species, as this designation would marginally constrain solar expansion.

Solar and wind farms are seldom close to the source of power consumption, requiring the construction of extensive transmission lines, which have proven to contribute to forest fires.

Solar is the fairest of the renewables in the eyes of most politicians, activists, and the news media.

The problem with solar is that the sun does not shine at night, and when it does shine during the day, it does not shine with the same intensity all day long or at every location. Even worse, on some days, it rains or it's cloudy.

Let's look at an ideal case, however, the case of California, with constant sunshine and blue skies in its vast deserts. As Albert Hammond sang, "it never rains in Southern California."

Solar has a serious problem with something known as the duck curve, a term coined in 2012 by the California Independent System Operator (CALISO).

CALISO's function is to manage the flow of power throughout the state, matching power production with power demand, on a minute-by-minute basis. In short, CALISO is literally responsible for keeping the lights on.

The electric power business is unlike any other businesses. Unlike Amazon, or your local grocery store, the power industry has no inventory.

There is no storage of electricity.

E. Joseph Duckett and Jeffrey L. Pierce

Power production must ramp up and ramp down immediately in response to changes in demand.

Figure 19

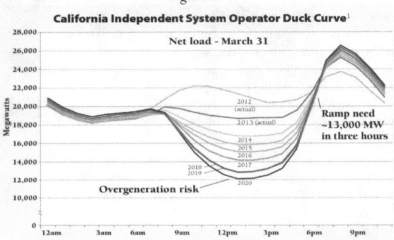

Source: D. Tait, Energy Alabama, 5/29/2017; Creative Commons license

The duck curve is illustrated on Figure 19. It shows total California power demand across a typical day, less the amount of power produced by solar.

As can be seen, the duck's belly grew, on that typical day, as more solar power came online between 2012 and 2020. Obviously, that's a good thing. California is making great progress in increasing the quantity of solar power the state produces and consumes. However, the duck curve lays bare a major technical problem with solar power.

Power demand peaks in the morning and peaks even more in the late evening. The peaks occur at the times when solar power is just coming back up to full power in the morning or when it is hardly available or not available in the evening.

In the evening, a massive ramp-up of natural gas fired power plants and/or import from adjacent Independent System Operators (if available) is necessary, to cover the loss of solar power. The belly will grow as California continues its aggressive policy toward expanded

114

renewable power production with policies that demand more electric power, such as mandating more electric vehicles and mandating reductions in natural gas use.

Is the duck curve a real problem? Yes, in August 2020, CALISO ordered rolling blackouts, while brief, in the evening, as predicted by the duck curve.

We close our discussion of renewable electric power with two further observations.

The first observation is that while it has been widely circulated that solar power generation has become cost-competitive with fossil fueled power generation, it has been overlooked that the power generating capacity supplying power, when solar is not generating power, is currently being provided by existing power generation capacity. When it is said that the cost of solar power generation is cost-competitive with fossil fuel power generation, the cost of solar power is not burdened with the indirect capital cost of these backup power generation facilities. At some point, additional natural gas fueled generation capacity will need to be constructed in parallel with solar generation capacity, or a technology must be developed to store solar power on a very, very large scale.

Reflecting the importance of backup (usually fossil-fueled) power, the National Bureau of Economic Research (NBER) has published a report entitled: *Bridging the Gap: Do Fast Reacting Fossil Technologies Facilitate Renewable Energy Diffusion?* [11]

It has been estimated that 8 Megawatts (MW) of back-up power are required for each 10 MW of wind or solar generation. Further, it has been estimated that storage capacity would be required to store at least three to five hours of solar-generated power.

At present, the only technology available to store electric power is batteries. However, the use of batteries to store large quantities of power is currently (pun intended) both technically and economically impractical. The NBER Report concludes:

> *...renewable energy integration has so far been possible thanks to the presence of fast-reacting back-up capacity based on fossil fuel technologies.*

The second observation is that policymakers and the news media seem to sometimes take inconsistent views of renewable electrical power. In July 2021, Governor Newsom of California wrote to Warren Buffet, whose Berkshire Hathaway owns the power company Pacific Corp., which owns hydroelectric plants feeding California with renewable energy. The Governor reiterated California's support for dismantling four hydroelectric dams located at the Oregon/California border. The dams were built between 1903 and 1964. They have been generating renewable power for decades. The reason Governor Newsom made this request was because it would improve salmon fishing for native Americans.

Understand that this would eliminate existing renewable power resources, resources that do run when the sun does not shine. We saw no published articles that mentioned any consideration of this tradeoff (renewable energy versus better fishing). Removal of the dams may have been the proper course of action, environmentally and/or socially, but it is surprising that this undesirable consequence was not being mentioned. The loss of this hydroelectric source must now be covered by natural gas fired generation. If we are battling an existential threat to mankind, what priority should have been given to restoring long lost fishing rights?

Obviously, this circles back to an earlier comment that renewable electric power is not without problems, and a later discussion that all relevant issues are not always vetted in the public eye.

Somewhat ironically, the most likely long-term solution to covering solar/wind power supply gaps and fluctuations is the same approach commonly used to adjust to demand fluctuations from nuclear/coal power—natural gas-fired power generation.[7] For decades, natural gas-fired generators, usually gas combustion turbines, have been constructed and operated as peak-shaving units and as backup units to nuclear and coal-fired generation units.

Typically, gas turbines can be started, stopped and modulated in much closer response to fluctuating energy demand than the coal or nuclear units. Even when the cost of natural gas was high, such turbines were used to supplement other cheaper power supplies (nuclear/coal) when excess demand required more power quickly.

In the case of renewable electric power, natural gas-fueled electric power might be considered a necessary prerequisite to assure a steady power supply in the event of sun/wind gaps. In a sense, natural gas, wind and solar may be viewed as complementary power generation technologies.

In fact, at least for the present, widespread renewable electric power production will require natural gas-fueled backups.

Fracking is a Net Environmental Problem

Hydraulic fracturing, or fracking, is the use of high-pressure water to improve the yield of oil and natural gas wells.

Fracking has been called "perhaps the most important energy discovery of the last half century." [12]

Figure 20
Schematic Geology of Natural Gas Resources

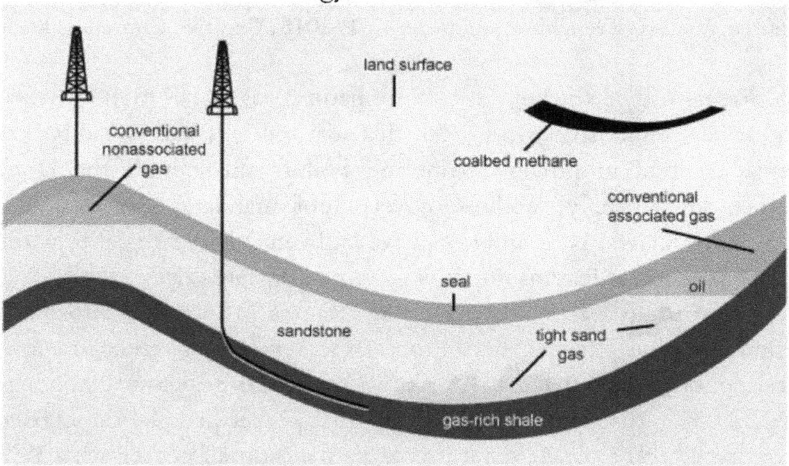

land surface

conventional
nonassociated
gas

coalbed methane

conventional
associated gas

seal

oil

sandstone

tight sand
gas

gas-rich shale

Source: Adapted from *United States Geological Survey factsheet 0113-01* (public domain)

Fracking has been opposed by some activist groups because of alleged negative impacts on air and water quality. Clearly fracking, like any mineral extraction process, poses environmental risks and must be properly performed.

On balance, however, fracking has produced a significant net benefit for the environment.[13]

Source: Victoria Trenchless Solutions, 6/1/2018; Creative Commons license

By dramatically reducing the costs of natural gas, it has made gas-firing so price-competitive with coal that natural gas has steadily been replacing coal in power plants and boilers throughout the United States, interestingly, without government mandates. Reductions in SO_2, particulates, NO_x and CO_2 have all been driven by coal to natural gas conversion at least as much as by regulatory pressures.

According to the United States Energy Information Administration, from 2010 to 2019, coal-fired electric power production dropped from 1,847,290 to 966,148 gigawatt hours per year. During that same period, natural gas power production increased from 997,697 to 1,558,815 gigawatt hours per year. While environmental regulations played some role in the shuttering of some coal-fired power plants, the economics of power production were the principal forces driving this shift.

The cost of natural gas dropped over 50% between the periods of 2005-2010 versus 2015-2020. As a result, it became less expensive to produce electric power with natural gas than with coal. When possible, power generators have permanently shut down coal-fired stations or have placed coal-fired stations in cold reserve. An efficient coal-fired power plant emits about 915 grams of CO_2 per kilowatt hour of power produced. A combined cycle natural gas fired power plant emits about 435 grams of CO_2 per kilowatt hour of power produced.[14] Fracking has made a major contribution to greenhouse gas control in the USA.

There are moves afoot to outlaw fracking. If fracking was suspended, natural gas supplies would tighten, natural gas prices would soar, and cold reserved coal-fired units would probably be brought back online, reversing the CO_2 reductions achieved. It does not seem like the anti-fracking lobby has considered this consequence. A recent study by the Energy Policy Institute at the University of Chicago has concluded that even for "the average household living in a community where fracking takes place, the benefits exceed the costs".[15] They noted that, on a local level, air pollution may increase due to increased truck traffic and diesel generators. They also noted that localized water contamination is probably <u>not</u> to blame for any health problems among people living near well sites.

Environment-Economy Conflict

When promoting new environmental initiatives, especially regulations, it is popular to describe them as win-win situations. Everyone wins. More regulation leads to both a cleaner environment and to improved economic conditions.

This win-win propaganda is Pollyannaish and illogical. Of course, environmental regulations requiring companies or municipalities to reduce their discharges are always costly. Few, if any, environmental improvements can be accomplished for free.

Of course, the costs of such environmental improvements may

Source: Australian Broadcasting Corporation, 2021; Creative Commons license

be justified based on reduced health risks or safer recreational waters, but such "wins" come at a cost. Ignoring costs is simply irresponsible.

At a 2013 American Association for the Advancement of Science (AAAS) Forum on Science and Technology Policy, the environmental regulation vs. economy issue was debated. It was noted that USA environmental regulations in the past three decades were shown to have resulted in approximately a 110% increase in imported goods from Mexico and Canada.[16] On a more local level, an estimated 500,000 jobs shifted from countries with plants that were <u>not</u> in compliance with stringent air quality standards to neighboring countries where less stringent standards could be met with no additional costs. Compared with total manufacturing job losses, a half-million job loss shift may seem small, but it is nevertheless a loss.

Source: jiffyavril, Freepik,2021; Creative Commons license

Offsetting economic and job losses, there have been undeniable improvements in air and water quality throughout the USA. The misconception is not that environmental regulations cannot yield environmental improvement but that such regulation does not carry significant costs. In some instances, maybe, but in general, regulation always adds cost.

---

[1] *Annual Report to the Nation: Cancer Death Rates Continue to Decline*, U.S. Center for Disease Control (CDC), March 2020

[2] U. of CA-Berkeley, *Clean Water Act Dramatically Cut Pollution in U.S. Waterways*, Science Daily, October 2018

[3] Air Quality Trends

[4] *State of the Air*, 2020, American Lung Association, April 2021

[5] Duckett, E. J., *Air Pollution in Allegheny County: Much Better Than You'd Think*, Letter to the Editor, Pittsburgh Post-Gazette, May 16, 2016

[6] *Emission Omissions: Lung Association Sees Only the Worst Case Scenario*, Editorial, Pittsburgh Post-Gazette, May 4, 2011

[7] 2019 Air Quality Annual Report, Allegheny County (PA) Health Department, April 2020

[8] Mosbergen, D., *China No Longer Wants Your Trash*, Huffington Post, January 24, 2018

[9] Roston, E., *Recycling: The Crisis After China's No*, Quick Take, December 2019

[10] *Future Energy: Report of the Energy Project at the Harvard Business School*, Edited by R. Stobaugh and D.Yergin, 1979

[11] Verdolini, E.,Vona, F., and Popp, D., *Bridging the Gap: Do Fast Reacting Fossil Technologies Facilitate Renewable Energy Diffusion?*, National Bureau of Economic Research, July 2016

[12] Barber, W., *Study Says Renewable Power Still Reliant on Backup from Natural Gas*, Power Engineering, August 2016

[13] Gold, R., *The Boom: How Fracking Ignites the American Energy Revolution and Changed the World*, Simon & Schuster, April 8, 2014

[14] *Reduced Emissions of CO_2, NOx and SOx from U.S. Power Plants Due to the Switch from Coal to Natural Gas with Combined Cycle Technology*, Chemical Sciences Division, NOAA Earth System Research Lab, Boulder, CO; J.A. de Gouw, et. al.; February 21, 2014

[15] Greenstone, M., *Fracking Has Its Costs and Benefits—The Trick is Balancing Them*, Energy Policy Institute at the University of Chicago, February 20, 2018

[16] Swan, N., *How Do Environmental Regulations Affect the Economy?*, Christian Science Monitor, May 12, 2013

E. Joseph Duckett and Jeffrey L. Pierce

CHAPTER 9: GLOBAL CLIMATE CHANGE

GLOBAL CLIMATE CHANGE is obviously a controversial and politically debated issue. A central question is how best to confront such change. An almost universally cited solution to confronting climate change is the adoption of alternative energy (i.e., energy that does not involve combustion and the release of CO_2) and shrinking carbon "footprints" as the way to arrest climate change. Viewed simplistically, this solution is offered because as the CO_2 concentration has risen in the atmosphere, global temperature has risen. There are reasons to question the practicality and effectiveness of focusing on promotion of alternative energy as the primary approach to addressing global warming.

Global Climate Change and Alternative Fuels

Although we are suggesting a different tact than most global climate change commentators, our approach is not about being climate change deniers. For at least the past 40 years, average global temperatures have been on the rise. Greenhouse gas (GHG) emissions, especially CO_2, have also been rising and the atmospheric concentration of CO_2 has increased. Combustion of fossil fuels is the major factor that has

HOLD IT!

driven the increased emission of CO_2

Figure 20

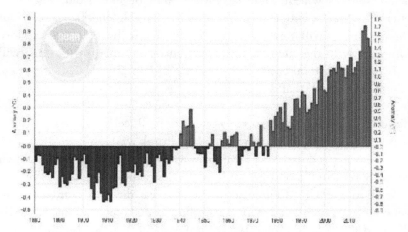

Departure of Global Temperature From Average, 1880 - 2018

Figure 21

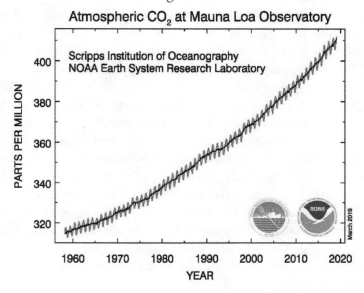

Atmospheric CO_2 at Mauna Loa Observatory

There is absolutely nothing wrong with encouraging solar, wind and hydroelectric power. There are reasons beyond global climate change to support fostering development of alternative energy sources. Long-term reliance on nonrenewable sources of energy does not make sense. These sources will at some far future point be exhausted. Improved energy efficiency extends the life of nonrenewable energy sources, but these sources will nevertheless still ultimately be exhausted. More importantly to the current discussion, improved energy efficiency will only defer CO_2 total emissions.

We suggest that waste-to-energy is another, but less obvious, legitimate alternative energy path, even though waste-to-energy directly or indirectly involves combustion of cellulosic waste (e.g., paper, food waste, wood waste, agricultural waste, manure) and its combustion produces CO_2. But this CO_2 is produced from carbon recently removed from the atmosphere. Its combustion can be viewed as carbon recycle. Cellulosic waste can be directly combusted to generate electricity or as a boiler fuel, or it can be anaerobically digested to produce renewable natural gas and then be combusted as a substitute for natural gas.

If cellulosic waste is landfilled, the decomposition of the wastes produces methane, another greenhouse gas. Methane is reported to be 25-30 times more potent as a GHG than CO_2. Environmental regulations now require capture of methane emissions from landfills. Landfill methane capture is generally assumed to be 75 percent, although its capture could be much higher. It is not possible to accurately measure its rate of capture. The captured landfill methane can be used for power generation, a direct natural gas substitute or for conversion to renewable natural gas. If the landfill methane is not beneficially used, it is flared. It should be noted that if the landfill methane is flared, its flaring is credited with prevention of methane emissions and is credited with conversion of methane to CO_2 (CO_2 being a lower strength GHG). Counterintuitively, combustion of landfill methane and the emission of CO_2 actually results in a significant GHG reduction. The beneficial use of landfill methane eliminates CO_2 emissions from the otherwise flared landfill methane and further mitigates the impact of landfill methane.

International Agreements

Much of the popular discussion of climate change is on the question of how best to address the problem. The current exclusive (or at least primary) focus is on adoption of alternative energy sources to prevent global climate change. There is reason to question whether the Kyoto, Paris and Glasgow Agreements—with their heavy reliance on voluntary international reduction of greenhouse gas (GHG) emissions—offer a realistic approach to arresting the potential effects of global climate change.

Hovering over the entire international agreement situation is that the buildup of GHGs is truly global.

Any country that takes its pledges seriously and dramatically converts its culture to non-fossil energy could find its own GHG emission reductions eclipsed by increases from other countries.

Any country that makes the first move to achieve zero greenhouse gas emissions risks that, unless all the other countries do likewise, their efforts could be both economically painful and climatologically pointless.

Being the first on the block to make the expensive switch to alternative energy could put a country at a serious competitive disadvantage economically—while potentially having no effect on net global CO_2 emissions—and hence no effect on actual global climate change.

Kyoto Protocol

Consider the recent history of international GHG reduction agreements. The Kyoto Protocol was signed in 1997 and set a series of "quantified emission limitations" for 39 countries.[1] Neither China nor India were included among the participating countries because they were classified as "developing" countries.[2]

The Protocol produced some peculiar results. First and most importantly, although the average target was a 5% <u>reduction</u> in GHG emissions by 2012, total worldwide emissions of CO_2 (considered the principal GHG) <u>increased</u> by approximately 19% between 2005 (when the Protocol was finally ratified) and 2012 (when the Protocol expired).[3] Also peculiar and counterintuitive is that the USA actually achieved its CO_2 reduction goals from the Protocol even though it never officially ratified the agreement.[4] Most of the USA reduction resulted from replacement of coal with natural gas, largely due to the cost-competitiveness of gas versus coal, as discussed in previous Chapters.

Figure 22
CO_2 Emission Trends by Country

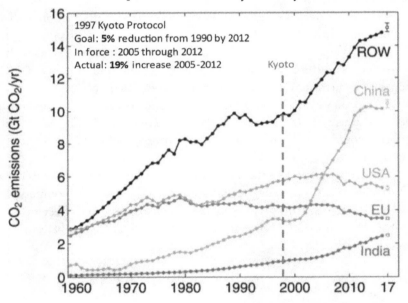

Source: Global Carbon Project; Creative Commons license

Paris and Glasgow Agreements

Skepticism about the Paris and Glasgow Agreements comes from three angles.

First is that the Paris Agreement itself (in Section II.17) admits that the temperature control objectives of the Agreement (1.5-2.0°C increase above pre-industrial global average temperatures) will not be met even if all of the pledged country-by-country emission reductions are achieved.[5]

Although not widely publicized, the Paris Agreement concedes that at least one more round of major pledges and reductions will be required to actually restrict global temperature increases to the goal of 2°C. In fact, some projections are that the temperature target will be missed by 3.5°C if only current pledges are met and that the current commitments would need to be tripled to achieve the 2°C targeted maximum increase.[6] China, currently the world's largest source of fossil fuel-based CO_2 emissions, has not set a quantitative upper limit but has pledged to peak their CO_2 emissions by 2030.

The second and more fundamental concern about these Agreements is that they are indefensibly Pollyannaish. Is there any example of serious international cooperation to solve a problem with long-term effects requiring difficult short-term solutions? Does anyone really believe that each country will honor and meet their pledged reductions without any enforcement provisions other than to reestablish new goals if the initial goals are not met? In particular, will China really restrict its greenhouse gas emissions after increasing CO_2 emissions more than threefold since 1997 (the year that the earlier global climate change agreement was signed in Kyoto)?

A third concern, especially applicable to the 2021 Glasgow Climate Change Conference, is the almost unanimous agreement that GHG emissions have continued to increase, and that none of the United Nation's three primary objectives for the Conference were met by the Glasgow Climate Pact. No commitments to cut global GHGs by 50% by 2030. No commitments to provide $100 billion/year of climate change financial aid. No commitments to assure that a full 50% of such financial aid would be applied to adaptation (rather than emission reduction). No major country, other than India, offered to revise its GHG reduction pledge from Paris, and it was reasserted that the current set of pledges won't achieve the stated climate temperature goals. In short, the Glasgow Conference was a bust.

There was only one significant agreement—to meet again in 2022 to set improved GHG targets!

Offsetting CO_2 Emissions

To illustrate the potential for reductions in one country to be offset by increases in another (hence, no net global reduction), consider China and the USA, the two largest CO_2 emission sources in the world. Although China is now the country with the largest total GHG emissions, the USA has a much higher GHG emission rate than China on a per-capita basis. So, considering only China and the USA mathematically, how much reduction in USA emissions would be needed to offset any per-capita rate increases in China?

Table 3—Calculated CO_2 Offsets—China versus USA

	China	USA
2018 CO_2 Emissions (10^6 Metric Tons/Year, MMTPY)	11,200	5,300
Population (10^6 people)	1427	327
Tons CO_2 per person (TPP)	7.8	16.2
What If?		
China TPP increases to ____% of USA	50%	
China's revised TPP	8.1	
Increase in MMTPY - China	364	
Required Offsetting USA reduction		7%
China TPP increases to ____% of USA	60%	
China's revised TPP	9.7	
Increase in MMTPY – China	2677	
Required Offsetting USA reduction		51%
China TPP increases to ____% of USA	70%	
China's revised TPP	11.3	
Increase in MMTPY – China	4990	
Required Offsetting USA reduction		94%

Assume: 2018 Actual CO_2 Emissions from fossil fuels
2018 Populations—no assumed changes
China's 2018 emission rate per person = about 48% of USA rate

As reflected in Table 3, based on 2018 data, the total annual CO_2 emissions from fossil fuels for China and the USA were 11,200 and 5,300 million metric tons (MMTPY) respectively. The corresponding annual per-capita emission rates were 7.8 tons per person (TPP) in China vs. 16.2 TPP in the USA.[7] Therefore, the TPP for China is only approximately 48% of that for the USA. TPP might be considered a rough surrogate for standard of living so it's entirely possible that China's TPP could increase as their country prospers economically.

Table 3 summarizes the emission changes that would result in a net zero balance between increases and decreases if China's TPP increases to approach that of the USA. As shown, a modest increase in China's TPP from 48% to 50% of the USA rate would require a 7% reduction in USA CO_2 emissions—just to offset the increase from China. Such offsetting would result in no net reduction in global CO_2 emissions. Steeper cuts would be required to achieve actual global reductions.

Similarly, an increase of CO_2 emissions from China by 2,677 MMTPY, corresponding to China's TPP increasing to only 60% of USA's, would require a 51% USA reduction. For perspective, between 2005 and 2017, fossil fuel-based CO_2 emissions from China increased by 4,614 MMTPY, approximately twice the amount of the calculated increase due to China's reaching 60% of the USA TPP.

Lastly, if China's TPP increased to only 70% of the USA rate, almost all (94%) of USA fossil fuel-based CO_2 emissions could be eliminated and there would still be no net reduction in worldwide CO_2.

There is evidence that coal-based CO_2 emissions from China are increasing.[8] Between 2000 and 2018, China tripled the amount of coal power it uses.[9] Also, China's transportation emissions of GHGs doubled between 2010 and 2020 (up 10 times from 1990 levels).

Another perspective on the tradeoff potential of USA-China CO_2 emissions compares China's announced pledges against reality. China has made two major pledges. First, that their CO_2 emissions will peak no later than 2030. Secondly, they have pledged to reduce their 2005 carbon "intensity" by 60-65% by 2030. Carbon intensity is the ratio of CO_2 emission tonnages to Gross Domestic Product (GDP), often measured in tons CO_2 per $1,000 GDP. It ratchets allowable CO_2 emissions based on economic conditions, rather than in absolute terms.

Summarizing a series of calculated CO_2 emissions based on China's pledges (see table below), if China's average GDP growth rate from 2018–2030 exceeds 2.5%, the increased CO_2 from China alone would offset all (100%) of the 2018 USA CO_2 emissions. This is even after a full 65% reduction in China's carbon intensity.

For reference, leaving aside the 2020 COVID-19 Pandemic Period, China's GDP growth rate has been 6-7%, well above the 2.5% rate calculated to <u>increase</u> China's CO_2 emissions above all potential USA CO_2 decreases.

USA-China Offsets:
Two Hypotheticals for Perspective

- If China's per capita CO_2 emissions increase to only **70%** of USA's (i.e. to 11.3 TPP), required USA reduction for **no** net global decrease would have to be 94% of USA's 2018 CO_2 emissions.

- If China's GDP growth rate from 2018-2030 exceeds **2.5%**, the increased CO_2 from China would more than offset 100% of all USA CO_2 emissions – even after a 65% reduction in China's Carbon Intensity
  - (China's growth rate over the past 10 years has been ~7%)

On a positive note, the rise in global CO_2 emissions will be slowed as alternative energy sources (even including natural gas) replace coal.

At least in part, this will be due to the public attention being drawn to climate change by the Paris and Glasgow Agreements and other initiatives.

Many people will respond to the calls for improved energy efficiency and less energy waste. But to be realistic, such public interest may only go so far towards the type of dramatic emission reductions reportedly needed to avert climate change.

To date, the willingness of the public to absorb increased costs, or to make any other personal sacrifices to combat climate change, has not been seriously tested.

HOLD IT!

Difficulties of Reducing CO_2 Emissions

Beyond the potential for increased CO_2 emissions from China and other countries, there are examples of backsliding on carbon reduction strategies even in countries that would seem to be in favor of GHG reduction. In France, the proposed expansion of its carbon tax triggered the 2018-2019 "yellow vest" protests and riots—and resulted in suspension of the tax increases.[10] In San Bernardino County, California, construction of large solar and wind farms on larger than one million acres of private land has been banned.[11]

As illustrated by the USA CO_2 reductions to date, displacement of coal with natural gas and other sources plays a large role in limiting CO_2 emissions. There are indications that complete displacement of coal is much more difficult than passing a law or signing an agreement. In 2001, the Canadian Province of Ontario announced that it would be closing the coal-fired 2,400 MW Lakeview Generating Station with the goal of eliminating all coal-fired electricity.[12] At the time, coal fueled only 25% of Ontario's total electric power supply so the objective seemed achievable. The initial coal phase-out target date was 2007. As 2007 approached, the target date was first extended to 2009 and then again to 2014. Today, there is no longer any coal-fired electric power generated in Ontario. Their power is now generated by nuclear (60%), hydro (24%), gas (9%) and renewables (7%). Not a mix likely to be duplicated soon by either the USA or China.

By the way, USA withdrawal or acceptance of the Paris Agreement is not all that pivotal to achieving the Paris Agreement goals. Unlike all other air pollution conditions in the USA, climate change is literally global, therefore requiring truly global cooperative solutions. USA CO_2 emissions have been on the decline since 2005, the reference year for the Paris Agreement. Over the period 2005-2017, USA CO_2 emissions were reduced by 14%.[13] Although not as steep a decline as written into the Paris Agreement, it is very possible that, largely through coal-to-gas conversion, the CO_2 reduction goals for the USA will be met with or without governmental endorsement of the Paris Agreement.

131

E. Joseph Duckett and Jeffrey L. Pierce

Montreal Protocol on CFCs

The previously mentioned "Montreal Protocol on Substances that Deplete the Ozone Layer (Chapter 5) could potentially provide an example of a successful international agreement addressing a global air quality issue, namely stratospheric ozone depletion. The Montreal Protocol succeeded in the elimination of the use of chlorofluorocarbons (CFCs). To date, the use of CFCs has been almost completely eliminated, suggesting a possible parallel with GHGs.

As reviewed in Chapter 5, however, for several reasons, the CFC experience is not easily repeatable for controlling GHGs. The use of CFCs was neither as prevalent nor connected to modern life as fossil fuels and CO_2 emissions are.

Adaptation

Two disturbing points were made above—1) prior international agreements, stretching back decades, have not been effective in slowing increases in CO_2 emissions, and that would cast doubt on the effectiveness of future agreements; and 2) the public's will to absorb the cost of elimination of CO_2 emissions has not been really tested. So where does this leave us?

Just as we take both preventive and corrective actions to address almost any problem (cancer, auto accidents, famine, poverty, or crime to cite a few examples), should policies focus our energies (pun intended!) on reactive/corrective treatments (termed "adaptation" measures in the Paris Agreement) at least as much as on prevention/mitigation. Reduction of GHGs by adoption of alternative energy sources and improvements in energy use efficiency are termed "mitigation" in the Paris Agreement. Most people wouldn't think of focusing solely on cancer prevention rather than seeking better treatments. Similarly, work to prevent traffic accidents doesn't preclude also preparing emergency services just in case.

Adaptation Plans

NYC Climate Resiliency Plan $19 Billion
- Building code revisions
- Staten Island sea wall

Source: M. Prieto, The Green Optimistic, 6/28/2019; Creative Commons license

If the climate models are to be believed, we are already experiencing some of the effects of global warming. Arctic ice shrinkage, violent storms and even forest fires have been linked to global climate warming.[14] Unless the Paris and Glasgow Agreement pledges will all be met and exceeded, modeled consequences should be expected to continue. If so, we should be preparing now to withstand them.

Improved and expanded agricultural irrigation methods, enhanced firefighting technologies, sea walls and even improved hurricane-resistant building construction should be prominently on our national—if not international—agenda at least as much as the encouragement of alternative energy sources. While on the subject of alternative energy—nuclear power (fission) could, at this point, be considered alternative energy. And what has happened to the prospect of nuclear fusion as a high-tech, zero-CO_2 source of electric power? As

133

Bill Gates notes in his 2021 book on climate change, quoting a nuclear science joke: "Fusion is 40 years away and it always will be." [15]

If we are serious about global climate change and the urgency of action, the time is right to better protect against the consequences of such change rather than to blindly assume that voluntary international global constraints on greenhouse gases will succeed. Several governments are already taking steps to adapt to climate changes.

New York City has adopted a "Resiliency Plan" to improve the city's capacity to withstand future extreme weather events. [16] The $19 billion plan, funded in 2019, includes changes in their building codes and construction of a 20-foot high Staten Island sea wall which would double as a scenic promenade. The Plan is an example of an adaptive action to prepare for possible climate-related emergencies. Likewise, flood control measures in the Netherlands, Venice and elsewhere may be considered "Defensive Design," a form of adaptation against potential climate change effects.

Another example of proposed adaptation is the San Antonio Climate Ready Program. [17] Under this program, San Antonio has set targets of achieving both carbon neutrality by 2050 and "communicating the necessity of adaptation." Specific adaptive projects are not identified in their Program, but at least it formally recognizes the important role of adaptation. The San Antonio Program cites examples of potential threats for which adaptive measures will be required: drought, flooding, vector-borne diseases and, of course, extreme heat. As a curious aside, San Antonio's largest sources of GHGs are landfills—rarely mentioned in GHG reduction plans and very difficult to design and manage for GHG capture and control.

Even the recently announced movement of the Indonesian capital from Jakarta to Borneo may be an example of adaptation to climate change. [18] One of the announced reasons for the relocation is to reduce the risks of floods and forest fires—two reported effects of global climate change. Back in the USA, the City of Buffalo, NY has announced a $30 million Environmental Impact Bond Program to, among other objectives, "help make our community more resilient to the impacts of climate change." [19]

The most recent (2021) USA federal plan for addressing the "Climate Crisis" included initiatives to strengthen adaptive measures across the country.[20] Among such measures are improved wildfire prevention and response.[21]

Probably at the extreme end of adaptive measures is the Pleistocene Park Project in Siberia recently described in National Geographic.[22] This Project aims to cool Arctic permafrost by returning Siberia to the Ice Age by, among other measures, reintroduction of herds of moose, bison, musk ox, yak and other temperature tolerant herbivores to revive the "Mammoth Steppe" ecosystem. The Project is described by its designers as an attempt to "defuse the carbon bomb".[23]

The need for serious attention to adaptive actions is already incorporated into the Paris Agreement but in a very under-publicized way. Within the Agreement, Article 9 adopts the Green Climate Fund as its financial mechanism, primarily to support projects in poor countries, which are disproportionately vulnerable to the effects of climate changes.[24] Included in this Fund authorization is an explicit provision that the funding of projects under the Green Climate Fund is to be split evenly (50/50) between "mitigation" and "adaptation" projects (paragraph 115). This is recognition that, especially for poor countries likely to feel the full brunt of climate changes, preparation for such changes is at least as important as global reduction of GHGs. Even for the USA, shouldn't we be bolstering our defenses against the potential effects of global climate change at least as much as promoting alternative energy?

Conclusion

In conclusion, if we are serious about global climate change, it is a mistake to think that it can be stopped by conversion to a carbon-free world economy. Repowering by adoption of alternative energy is both technically achievable and environmentally—sometimes even economically—beneficial for local areas. However, such extensive repowering is unlikely to succeed worldwide, at least not on the scale and schedule sufficient to arrest the potential effects of climate change. Adoption of adaptation is advisable and, in fact, necessary.

[1] Clark, D., *Has the Kyoto Protocol Made Any Difference to Carbon Emissions?*, The Guardian, November 2012

[2] Rapier, R., *Global Carbon Dioxide Emissions Set New Record*, Forbes, June 2018

[3] *US Energy Related Carbon Dioxide Emissions 2017*, US Energy Information Administration, September 28, 2018

[4] Frolich, T.C., 25 *Countries That Produce Most of the CO_2 Emissions*, 24/7 Wall Street Special Report, June 5, 2019

[5] Harvey, C., *CO_2 Emissions Reached an All-Time High in 2018*, Scientific American E&E News, Dec 6, 2018

[6] *Kyoto Protocol to the United Nations Framework Convention on Climate Change*, United Nations, 1998, Annex B

[7] Heubi,B., *Is China Returning to Coal-Fired Power?*, Engineering & Technology, March 28, 2019

[8] Myllyvirta,L., *China's CO_2 Emissions from Fossil Fuels and Cement Production Grew by an Estimated 4% in the first half of 2019*, Carbon Brief, May 9, 2019

[9] Gates, B., *How to Avoid Climate Disaster*, Alfred A. Knopf, New York, New York, 2021

[10] Rampell, C., *Learning from France's Flubs on Climate*, Washington Post, June 21, 2019

[11] Roth, S., *California's San Bernardino County Slams the Brakes on Big Solar Projects*, Los Angeles Times, February 28, 2019

[12] Harris, M., Beck, M., and Gerasimchuck, I., *The End of Coal: Ontario's Coal Phase-Out*, International Institute for Sustainable Development, June 2015

[13] Bailey, R., *Carbon Dioxide: US Emissions Down; European Emissions Up*, Reason, May 2018

[14] Lustgarten, A., *When States Are No Longer Habitable*, New York Times Magazine, October 2, 2020

[15] Gates, B., *How to Avoid Climate Disaster*, Alfred A. Knopf, New York, New York, 2021

[16] Freedman, A., *New York Launches $19.5 Billion Climate Resiliency Plan*, Climate Central, June 11,2013

[17] Murphy, J., *SA Climate Ready*, Environmental Management (EM), September 2019

---

[18] *Indonesia to Move Capital from Sinking Jakarta to Borneo*, Associated Press, September 1, 2019

[19] *City of Buffalo to Launch the Largest Environmental Impact Bond in the Country*, WaterWorld, February 7, 2020

[20] *Biden Releases Plan to Tackle Climate Crisis*, Coatings Industry News, November 24, 2020

[21] Kaplan, S., *U.S. Unready for Wildfire Escalation*, Washington Post, July 2, 2021

[22] Welsh, C., *The Threat Below*, National Geographic, September 2019, pp 74-79

[23] Worrall, S., *We Could Resurrect the Wooly Mammoth—Here's How*, National Geographic, July 10, 2010

[24] *Green Climate Fund and the Paris Agreement*, Climate Focus, February 2016

E. Joseph Duckett and Jeffrey L. Pierce

CHAPTER 10: SOME SOLUTIONS

STEERING TOWARD TRUTH and sound environmental decision-making has always been tough. The sheer volume of data, analyses and biased "spins" makes it even tougher. There are, however, some approaches that could work.

Environmental Dispatching

Some environmental issues develop over decade-long cycles of sources and effects. For such issues (e.g., global climate change or oceanic mercury buildup) short-term blips in emissions or environmental concentrations are essentially irrelevant. Only the long-term trends matter. These are measured as averages, often over years or even decades.

But there is another class of environmental concerns that have more acute effects and therefore are tracked on a much shorter-term basis. One of the most infamous of such short-term episodes was the 1948 air pollution disaster in Donora,

138

PA. Under conditions of heavy industrial emissions and a prolonged atmospheric inversion, 20 people died in only 5 days. Obviously, a dramatic episode of acute health effects from environmental releases.[1]

One of the approaches that can be used to avert—or at least lessen—the impact of short-term episodes is "environmental dispatching." This term applies primarily to large pollutant dischargers like industrial facilities and power plants.

Source: Y.Lambrev, Wikimedia, 2/8/2009; Creative Commons license

Figure 23
PJM Dispatch Pattern

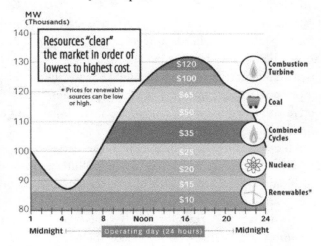

Source: PJM Learning Center, 2018; Creative Commons license

E. Joseph Duckett and Jeffrey L. Pierce

The idea is to adjust short term emissions (which may include wastewater discharges) when short-term weather conditions would heighten the impact of these releases. For example, a dual-fueled boiler (say, coal and natural gas) could be required either to reduce production or switch to a lesser polluting fuel when an atmospheric inversion is forecast. Similarly, sources that emit VOCs or NOx, the precursors of ozone, could be required to reduce emissions (including by reducing production) in anticipation of hot steamy weather—the conditions that favor ozone formation.

Varying production rates in response to fluctuating energy demand is practiced all the time by the electric utility industry. It is known as "dispatching." It is intended to supply electricity to satisfy short-term energy demand as quickly as possible and at the lowest possible cost of generation. Under dispatching arrangements, the lowest cost generators are given first priority. When demand exceeds the capacity of the lowest cost plants, the next lowest cost and most responsive units are brought online. The highest priced generators are only activated when needed to satisfy peak demand.

In ordinary electric power dispatching plans, the priority for activating any individual plant/unit is based on its economic cost of production. When demand shoots up, the speed with which a power plant can ramp up production also becomes a big factor. Dispatching is aimed at achieving the most cost-effective and responsive mix of generators. Natural gas-fired power plants are typically the go-to generators when demand surges occur.

Environmental dispatching would extend the concept beyond production/generation costs to include pollution rate.[2] From the perspective of environmental protection, such dispatching could avert high impact pollution effects. It would be a form of peak-shaving to reduce high concentrations of pollutants rather than electricity demand peaks. It would be most applicable to air pollution episodes but may also possibly be useful in the event of a water supply emergency, like a breakdown in a potable water supply plant. As with conventional power plant dispatching, the most likely sources to be activated during an air pollution emergency would probably be natural gas-fired.

From the perspective of an industry or utility, the advantage of adopting environmental dispatching as a regulatory policy would be that there could be some relaxation of emission requirements when unnecessary to control short-term health effects. Obviously, capital investment costs cannot change based on a short-term situation. But some environmentally related operating costs, such as ammonia injection for NO_x control, can be adjusted—up or down—based on short term air quality conditions. To some extent, the seasonally varying controls for VOCs and NO_x to restrict ozone formation are already a form of environmental dispatch based on the warmer months of the so-called "ozone season." But, environmental dispatching, as suggested here, would be aimed at much shorter and more temporary air quality episodes than entire multi-month seasons.

Continuous Prioritization

Bureaucracies are notorious for their self-preservation instincts. Organizations, both in government and within corporations, are often created in response to a problem. Even after the problem has been resolved, however, these same organizations often resist disbandment. Thus continues the cycle of bureaucratic growth and scope expansion.

In the environmental field, it is very rare for any regulatory agency— or corporate office—to admit that their purpose has been served and that they can be either dissolved or reabsorbed into their organizations.

This self-protective instinct is understandable, but it leads to uncontrolled scope creep. It also limits the staff and funds available to address newly emerging environmental issues.

Source: N. Youngson, Alpha Stock Images, 2020; Creative Commons license

One remedy for the expansive upward-ratcheting of environmental organizations is insistence on periodic reevaluation of priorities. All too often, whenever reevaluations are performed, they result in <u>adding</u> new priorities without eliminating or even deemphasizing any.

A reflection of the reluctance to prioritize environmental issues is the oft-repeated response of "everything is a priority" when asked about ranking of environmental issues. If everything is a priority, then nothing is a priority.

Prioritization among environmental programs would require serious thought about the criteria to be assessed. Are human health effects more important than threats to aquatic ecosystems? Are toxicities of materials more important than their routes of exposure? Are childhood health effects ranked above adult risks? Are short-term/acute effects more important than long-term/chronic?

Several years ago, the Pennsylvania Department of Environmental Protection attempted to identify top priorities for its programs. After harvesting ideas from a wide range of sources, the Department published a list of more than 300 "priorities." noting that they were not being ranked.

Establishing a huge number of unranked priorities was not priority-setting at all. Instead, it was a simple recital of ideas and problem areas. Real prioritization would require a rigorous ranking of issues and objectives. It would also require identifying some issues which are <u>not</u> high priorities and can be deemphasized. This would undoubtedly lead to objections from advocates for the deemphasized issues. Prioritization requires decision-making, often at the risk of dissention. Continuing to function without strong prioritization is wasteful, dysfunctional, and unfortunately all too often, rampant.

Prevention vs. Responsiveness

It is obvious that preventing a problem—almost any problem—is preferable to reacting to the problem after it has occurred. The trick is to figure out preventive measures and harness them efficiently.

Despite lip service to prevention, some of the most impactful and measurable environmental activities are in response to unprecedented episodes. Hurricanes, floods, spills, fires, accidents and atmospheric inversions all result in immediate risks to human health and the environment.

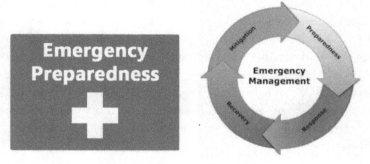

Source: US NASA,2020; Creative Commons license

To the extent that popular consensus ever coalesces around environmental issues, it is most often during—and in response to—major dramatic events. Hurricanes and floods jeopardize public water supplies, thus immediately threatening human health. Similarly, major fires, including forest fires and industrial explosions pose serious inhalation threats.

During such events, when the threats of health damage are immediate and severe, the spotlight turns to environmental authorities to quickly assess risks, take remedial actions and minimize health impacts.

In the absence of immediate threats, any consensus on environmental issues is much harder to find. Slow-moving crises or distant potential risks don't draw the same urgency. Prevention is therefore a tougher "sell" than responsiveness.

Foresight is never 20/20 and there is often distrust of dire predictions requiring costly and urgent actions, justified in the name of prevention. Few people agree that prevention isn't preferable to correction but, practically speaking, prevention doesn't always work. Environmental damage episodes can occur despite attempts to prevent them.

So, given the attention paid to environmental episodes and the immediacy of response required, why not deliberately strengthen the responsiveness of environmental organizations. This could apply both to governmental agencies and private businesses.

An emphasis on responsiveness would require much more than writing procedures and response plans to fill bookshelves. These would need to be accompanied by periodic "mock" response drills, strengthened detection systems and stockpiles of key response supplies/equipment. Such advance response planning is routine for military, police and medical organizations but less routine for environmental organizations. Yet, some of the most effective impacts on environmental threats to health can be accomplished by responsiveness.

Cross-Boundary Regulation

At the risk of Pollyannaism and hopeless positivity, it seems obvious that many of today's most thorny environmental issues cannot be addressed within political boundaries. Examples include: mercury contamination; acid precipitation (acid rain); oceanic plastic pollution and, of course, climate change.

In each of these cases, pollutants and effects transcend the borders of states, countries and even continents. There are few, if any, examples of successful trans-country environmental control programs. As noted in Chapters 5 and 9, the Montreal Protocol on Substances that Deplete the Ozone Layer may provide one successful example but as discussed, CFC control was more manageable than most other global—or even regional—environmental problems.[3]

Cross-National Environmental Protection ?

Source: pixabay, 5/25/2014;
Creative Commons license

Global
Cooperation
An Uphill Battle

Source: International Moinetary Fund,2016; Creative Commons license

Although much easier said than done, cross-boundary cooperation in reining in serious environmental risks will require diplomacy, creativity, trust and popular support. To enlist the involvement of poorer countries, transfers of money will be needed. Not least of all, it may also require some luck!

Informed Consumer Decision-Making

In the normal course of reviewing environmental protection options, the emphasis is usually on governmental or corporate actions such as regulations, subsidies, process changes and material substitutions. Often ignored or overlooked are the choices made by individual consumers as they purchase and use products.

As discussed in Chapter 7, there is no shortage of bogus attempts to influence consumer behavior by appealing to questionable environmental claims. Gimmickry does not relieve any real environmental risks.

In contrast to misleading marketing promotions, there are real and significant environmental implications of many individual consumer decisions. An eye-opening review of such decisions was published as "The Consumer's Guide to Effective Environmental Choices" by the Union of Concerned Scientists.[4] This Guide rated 50

145

categories of consumer items, based on their environmental impact. Among these, the top 7 accounted for the large majority of impacts.

This Guide detailed a wide variety of both significant and insignificant decisions with environmental implications. It pointed out that the purchase of a car, for example, has much more impact than purchasing such environmentally controversial products as disposable diapers. In addition to automobiles, the highest ranked impacts included home heating, air conditioning and household appliances. Microwave ovens, for example, use only one-third of the electricity of an electric oven. The lowest impact items included spray cans and paper napkins. Some choices were judged as individually high impact but only a small share of consumer spending. These included household pesticides, fertilizers, fireplaces and lawnmowers. The Guide was published in 1999. If written or updated today, it would no doubt list plastic straw bans as among the insignificant actions.

To make informed environmental choices, consumers need several things. First, they must have the inclination to reduce pollution and/or conserve resources. Secondly, they must have environmentally advantageous options available. Thirdly, and most importantly, consumers need understandable, reliable and authentic information on the choices available.

In short, an environmentally responsible consumer must be an informed consumer. This is much easier said than done. Who would set the definitions and measurements of environmental impact? How would potential impact(s) be prioritized, ranked and weighted? Who would authenticate the measured impacts? Above all, how much effort or cost would be required from consumers to make the "right" choice? As noted in the above-referenced Guide, individual consumer action works best when it does not require significant consumer sacrifice.

There are some noteworthy precedents for evaluating consumer products. Consumers Union, for example, regularly measures and rates the characteristics of a wide variety of products. Few such evaluations, however, single out environmental impacts as prime criteria for rankings.

Realistically, the best approach to assisting consumers in making informed environmental choices may be by narrowing the impact measures to only one or two criteria. There would be argument over which criteria to select but the strongest argument might be in favor of simplification by focusing on the most important impacts. Fuel economy ratings for cars, power consumption ratings for appliances and even water consumption data for toilets are all examples of environmentally-related measures for consumer products.

What's missing from these existing ratings are subtleties. Such criteria as reliance on scarce and/or toxic raw materials, disposal restrictions or manufacturing power consumption all affect the overall impact of products. Incorporating such subtle measures may overcomplicate environmental "ratings" and would certainly introduce difficulties but should not be completely dismissed. Except for hazardous wastes, even the U.S. EPA has no real definition of "recyclable" for consumer products.

One particularly big area for which there is little environmental clarity is consumer packaging. If it was not evident before, the food takeout surge and emphasis on cleanliness during the COVID-19 crisis has launched an expansion of concern for the uptick in single-use containers. It is estimated that the consumption of single-use plastics has increased by 250-300% in 2020 due to the COVID-19 Pandemic.[5] Would it be instructive to distinguish Styrofoam containers (generally not recyclable) from paper-based (at least theoretically recyclable) boxes?

The most likely successful approach may be to simply differentiate the "big" environmental choices from those with little or no impact. Returning to the above-referenced Guidebook, maybe it's time to revisit and evaluate choices among today's array of products—from electric cars to cell phones to genetically modified foods.

Market-Based Air Emissions Regulation: Cap-and-Trade

For several decades, economists have hypothesized that using market-based mechanisms to restrict emissions can achieve air quality goals with maximum economic efficiency. The most prominent suggested market approach is the use of so-called "cap-and-trade" regulations.

Most air emission limits in the USA take the form of specific "command and control" restrictions on individual sources, expressed either in standard regulatory provisions or in individual permit conditions. Under a cap-and-trade approach, sources have the option of either restricting their own emissions or purchasing emission "credits" from another source that has over-controlled theirs.

Under cap-and-trade, sources are annually allocated a permissible maximum quantity of emission allowances. The total allowance pool, usually set by state regulatory agencies, is based on requirements to meet National Ambient Air Quality Standards (NAAQS). The distribution of allowances (called an "allocation") among sources is based on their relative size and emission potential. If allowable ambient concentrations are reduced via revisions to the NAAQS standards, the caps are correspondingly lowered and, in turn so are the allowance allocations.[6] In theory, as long as the caps are met, ambient air quality should be protected.

At the end of each year, a source's actual emissions are compared against their allocation. If a source has not used all its allocation, it has achieved compliance and has excess allowance credits which can either be "banked" for their own future use or sold to another source that had emitted more than they had been allowed (that is, exceeded their allowance). The cap-and-trade system is very similar to monetary banking. In fact, the repository for unused allocations is often referenced as an emission allowance bank.

For individual sources, cap-and-trade provides a strong economic incentive to keep emissions under the allocations so that the excess can be sold or saved for the future. The incentive includes not having to purchase costly allowances from another source. For sources that over-emit their allowances, it provides a way to supplement their allocation and remain in compliance. If the market purchase price for unused allocations is above the costs of control, the economically smart decision for an individual source is to control their own emissions rather than having to buy allowances from the market.

Especially in the case of SO_2 controls (see Chapter 3) and to a lesser extent for NO_x and VOC controls, the cap-and-trade approach has been very successful both in keeping total emissions within the established overall caps and in doing so in a cost-efficient way. This approach has also been suggested for mercury control but has not yet actually been adopted (see Chapter 6). Most recently, this approach has also been proposed for reduction of CO_2 emissions.

A frequently cited argument in favor of command-and-control and against the use of cap-and-trade controls is the potential creation of air pollution "hot spots." These are local areas with excessively high pollutant concentrations. At least in theory, such unhealthy localized concentrations can unintentionally result from the clustering of similar sources. Maintaining a regional cap on emissions may not guarantee that a local area does not become a hot spot. By its nature, the cap-and-trade approach focuses less on the location of specific emission sources than on the quantity of pollutants emitted over a large area. The "cap" is an expression of a regional (or even nationwide) emission limit.

Despite speculation that the use of cap-and-trade regulation leads to the creation of hotspots, there is little or no evidence to support this. Cap-and-trade programs for SO_2 have been in place for many years with success in reducing area-wide ambient levels but no indications of hot spots.[7]

The linkage between cap-and-trade regulation and hot spot creation was the subject of a Chicago area study.[8] After adoption of a cap-and-trade program for stationary sources of VOCs, 89 out of 95 areas (defined by zip codes in this study) experienced a decrease in VOC emissions while only 6 experienced an increase. When the areas were expanded slightly (adding adjacent zip codes), all the areas showed decreased VOC emissions. The areas with the largest initial VOC emissions experienced the most significant reductions after trading. In other words, the authors concluded that adopting the cap-and-trade approach not only did not increase the concentration of source emissions (i.e., create hotspots) but tended to both reduce ambient concentrations and spread the distribution of emissions more widely across the affected region.

There is a plausible explanation for broadening, rather than concentrating, the sources of emissions after adoption of a cap-and-trade program. Usually, it is more economical, on a cost per ton basis, to control large emission sources than smaller ones. Cap-and-trade regulations are based on a quantitative aggregate emission cap. Due to economies of scale, it is almost always more economical to reduce a ton of emissions from a large source than to achieve the same reduction from a smaller one.

Our bottom-line conclusion is that adoption of a cap-and-trade regulatory approach is at least as likely to dissipate as to create hot spots, and further, that offering market-based incentives to regulate emissions makes a lot of sense.

Beware of Narrative Control

An underlying theme in this book is to be careful of accepting what you are being told as fact, or accepting what you are being told, as the full story. The following is attributed to Buddha:

> *Do not believe in anything simply because you have heard it. Do not believe in anything simply because it is spoken and rumored by many...But, after observation and analysis, when you find anything that agrees with reason and is conducive to the good and benefit of all, then accept it...*

While things were a lot simpler in the 4th century BC, Buddha's words provide good counsel today. Many issues requiring decisions today are more technically complex than most of the public has the time to fully research or the background to fully understand. We hope that this book illustrates the need for individuals in positions of authority, and for the general public, to at least attempt to learn something, beyond the superficial, about an issue before jumping to conclusions, and supporting a position or a cause. In other words—Hold It!

Source: M. Wiggins,2/22/2018; Creative Commons license

In almost all situations, there is a truth or truths. Parties who want you to agree with them may "tell the truth" as they see it, but they often leave out critical information and misrepresent the facts, either intentionally or unintentionally. The process of conveying incomplete truths to others is called "narrative control." If done unintentionally, John Kenneth Galbraith might have called this another example of "innocent fraud." However, when control of the narrative is undertaken to serve an entity's financial interest, or to increase a group's political power, such fraud is anything but innocent.

Relentless Search for Truth

As clichéd as it sounds, the most difficult task in the field of environmental protection, is the search for truth. This is not unique to the environmental realm, but it certainly applies to almost every current environmental and/or health issue.

In past decades, reliance on refereed journals and peer-reviewed scientific reports was sufficient to illuminate environmental—and other—truths. Granted, there has always been disagreement and debate about scientific methods and conclusions. But if the debate is conducted under established procedures and with rational reasoning, there are guardrails that at least protect the credibility of environmental information.

Today, the Internet firehose of information routinely presents a wide variety of environmental data and opinions, often without authentication. Complex phenomena are too often compressed into catchy shorthand phrases: Climate Change; Air/Water Pollution; Wasteful Packaging, etc., with little acknowledgement of nuances and subtleties. As if you can choose whether you are for or against Climate Change!

As discussed throughout this book, environmental issues are rarely simple. So, to properly address these issues requires a thorough understanding and appreciation of complexities, tradeoffs, efficiencies, and costs.

Maybe the single biggest challenge for the next generation of environmental professionals is to identify reliable sources of valid information. Put simply, this translates into searching for the truth. The search requires intellectual curiosity, critical thinking, and relentless research. It requires intellectual humility. It also requires the courage to go where the facts lead, irrespective of preconceptions and conventional wisdom. It can be frustrating, exhausting and confusing, all at once. As one writer has recently put it:

"The search for truth is bumpy and complicated...we should be cautious about shoehorning facts into our ideological constructs." [9]

Ironically, those who question conventional wisdom about environmental issues are often tagged as wishy-washy equivocators who resist adopting simplistic advocacy positions—whether alarmist or dismissive. Accepting easy theories and solutions can be attractive traps.

As difficult as it is, the search for truth takes both perseverance and courage. Without such searching, environmental protection will be jeopardized. With it, intelligent progress will proceed. Searching for truth before jumping to environmental conclusions often takes time and patience. In short, when faced with pressure to make jumpy simplistic decisions, it's usually best to say, "Hold It!"

HOLD IT!

[1] Snyder, L.P., *Revisiting Donora, Pennsylvania's 1948 Air Pollution Disaster*, Devastation and Renewal, J.A. Tarr (ed.) 2003

[2] Wu, L.H. Wang, Y.N., et al, *Environmental/Economic Power Dispatch Using Multi-Objective Differential Evolution Algorithm*, Electric Power Systems Research, September 2010

[3] Rae, I., *Saving the Ozone Layer: Why the Montreal Protocol Worked*, The Conversation, September 9, 2012

[4] Brower, M. and Leon, W., *The Consumer's Guide to Effective Environmental Choices*, Union of Concerned Scientists, Three Rivers Press, NY, NY, 1999

[5] Zimmermann, S., *Plastic Waste Problem Amplified by the Pandemic*, Chicago Sun Times, November 11, 2020

[6] *Sulfur Dioxide Emissions from U.S. Power Plants have Fallen Faster than Coal Generation*, Today in Energy, U.S. Energy Information Administration, February 3, 2017

[7] Duckett, E.J., *Targeting Air Emissions: Are There Hot Spots?*, Proceeding of the Air& Waste Management Annual Conference and Exhibit June 2011 (Paper #2011)

[8] Kosobud, R., Stokes, H., Tallarico, C., *Does Emission Trading Lead to Air Pollution Hot Spots?*, International Journal of Environmental Technology and Management, Volume 4, Nos 1-2 (2004) pp. 137-156

[9] Kristof, N., *How to Reach People Who Are Wrong*, Pittsburgh Post-Gazette, March 11,2021

E. Joseph Duckett and Jeffrey L. Pierce

ACKNOWLEDGEMENTS

THE AUTHORS ARE solely responsible for the observations made in this book, and for selection of the materials, which the authors hope sufficiently illustrate the authors' premise.

The authors do, however, want to specifically acknowledge the assistance of two individuals, without whom this book would not have reached a publishable form. First, Joanne Cosgrove, who converted hundreds of pages of Joe's scribbled notes, loose images and tables into a first draft.

Second, Jennifer Mancini, Jeff's administrative assistant of over 20 years, who converted his scribbled notes; and reformatted the text, figures and tables in this book to comply with the publisher's requirements.

INDEX

ABOUT THE AUTHORS

E. JOSEPH (JOE) DUCKETT PH.D., P.E. has more than 50 years of experience in environmental regulation, research, engineering, management and teaching. He has a Doctorate in Environmental Engineering from Drexel University and Masters Degrees in Environmental Health (University of Pittsburgh) and Business Administration (Drexel). He is a registered Professional Engineer in 6 states and has authored and/or presented more than 90 publications.

He has been an active member, and often committee chair, of the Air & Waste Management Association, American Society for Testing & Materials and the Engineers Society of Western Pennsylvania. He is an appointed member and past chair of Pennsylvania's Air Quality Technical Advisory Committee.

He has taught environmental health and emission control for the past 25 years and is currently an adjunct faculty member within the University of Pittsburgh's Swanson School of Engineering.

He and his wife Joellen live in Pittsburgh PA, with frequent children and grandchildren visits.

JEFFREY L. PIERCE, P.E. has almost 50 years of experience in environmental and renewable energy engineering. He has a Bachelors Degree in Civil Engineering and a Masters Degree in Environmental Engineering from the University of Pittsburgh. He is a registered Professional Engineer in Massachusetts and California. He has made more than 100 presentations on environmental and energy topics at technical conferences in the United States and Asia.

He played a major role in the implementation and design of the world's largest landfill gas fueled power plant (1986), the world's largest coal mine methane power plant (2008), and the world's largest renewable natural gas plant (2021). The aggregate total of these, and other waste methane utilization projects, have avoided in excess of the equivalent of an estimated 100 million metric tons of carbon dioxide emissions from the atmosphere.

He currently resides in Torrance, CA, after relocating from Pittsburgh, PA in the mid-1980's.

CPSIA information can be obtained
at www.ICGtesting.com
Printed in the USA
BVHW061604150322
631522BV00007B/557